CONSTRUCTION INDUSTRY TRAIN

BS 7671: Requirements for Electrical Installation

IEE Regulations

(16TH EDITION)

STUDY NOTES

Revised February 1993

Published by:-
*Construction Industry Training Board,
First Edition, 1982*

*This edition published 1992
Revised February 1993*

ISBN 1 85751 045 3
© Construction Industry Training Board

PREFACE

This publication retains the presentation style of the previous Study Notes for the 15th Edition of the Wiring Regulations which sought to reconcile design requirements for electrical installations, with the actual work practices and methods employed by practising installation electricians.

In respect of differences between the 15th and 16th Editions, the introduction of amendments and new information is denoted in this publication by the use of the symbol ⟨✳⟩ in the margin.

Additional and updated illustrations and worked examples have been included in this revised edition of the Study Notes, together with project assignments.

The CITB wishes to express its thanks to **Astra Training Services** and **JT Ltd. (JTL)** for their assistance in preparing the revised material contained in this publication.

NOTE:

These Study Notes contain abbreviated extracts and paraphrases of the IEE Regulations. It is emphasised that these interpretations of the Regulations have been devised for the purpose of training and should not be regarded as authoritative in any other context. When necessary, the Regulations should be referred to directly.

CONTENTS

ACKNOWLEDGEMENTS

The Construction Industry Training Board wishes to express its thanks to the following organisations for their co-operation in allowing extracts and illustrations from their various publications to be reproduced in these Study Notes:

The Institution of Electrical Engineers

15th Edition of the IEE Wiring Regulations

Commentary on the 15th Edition of the IEE Wiring Regulations

A Guide to the 15th Edition of the IEE Wiring Regulations

The Electrical Contractors Association

The Electrical Contractors Association of Scotland

The National Inspection Council for Electrical Installation Contracting

A Handbook on the Fifteenth Edition of the IEE Regulations for Electrical Installation

The British Standards Institution

Astra Training Services

We also wish to thank the following companies for permission to illustrate certain of their products:

Crabtree Electrical Industries Ltd.

Evershed and Vignoles Ltd.

G.E.C. Fusegear Ltd.

W.J. Furse & Co. Ltd.

Davis Trunking Ltd.

BICC Components Ltd.

BICC Pyrotenax Ltd.

Clare Instruments Ltd.

Symbols and Formulae

The following symbols and formulae are employed in the various modules.

Module 3 - Symbols

U_0 — nominal voltage to earth in volts

I_p — prospective short circuit current in amperes

Z_E — earth fault loop impedance external to that part of a circuit in ohms

f — frequency in Hz (Hertz - cycles per second)

Module 5 - Formulae

Assessed current demand for an electric cooker:
The first 10A of total rated current of cooker + 30% of the remainder of the total rated current + 5A if a socket outlet is incorporated in the control unit.
Assessed current demand = 10A + 30% of remainder + 5A

Discharge lighting current demand:
Where exact information is not available and the power factor is assumed to be less than 0.85.
Current demand of a discharge lamp

$$= \frac{\text{wattage of lamp} \times 1.8}{\text{supply voltage}}$$

$$= \frac{W \times 1.8}{U_0}$$

Module 6 - Symbols

Z — impedance (ohms)

R — resistance (ohms)

Ω — ohm; $k\Omega$ - kilohm (1000 ohms);
$M\Omega$ - megohms (1,000,000 ohms)

Z_s — earth fault loop impedance of the system

R_1 — resistance of phase conductor from origin of circuit to the most distant socket outlet or other point of utilisation. Z_1 for conductors > 35 mm^2

R_2 — resistance of protective conductor from origin of circuit to the most distant socket outlet or other point of utilisation. Z_2 for conductors > 35 mm^2

< — less than $\leq$ — equal or less than

> — greater than $\geq$ — equal to or greater than

$\simeq$ — approximately equal to

cpc — circuit protective conductor

I_f — earth fault current in amperes

Module 6 - Formulae

$$Z_S = Z_E + (R_1 + R_2)$$

$$I_f = \frac{U_O}{Z_S}$$

Module 7 - Symbols

Ib	—	design current of the circuit in amperes
In	—	nominal current or current setting of device
Iz	—	effective current carrying capacity of any of the circuit conductors
I2	—	effective operating current of a device
t	—	duration in seconds
k	—	a constant value for a particular type of cable
S	—	cross-sectional area (csa) of conductor in mm^2
I	—	effective short circuit current in amperes
It	—	tabulated value of current in a circuit in amperes

Module 7 - Formulae

Prospective short circuit current Ip

$$I_p = \frac{V}{Z_t + Z_1 + Z_2}$$

Where V = source voltage

Z_t = impedance of supply transformer

Z_1, Z_2 = conductor impedances ($Z_2 = Z_n$ for single phase and neutral circuit)

Protection against overcurrent

In = nominal current or current setting of device

Ib = design current of circuit

Iz = current carrying capacity of any of the circuit conductors

I2 = current which ensures effective operation of the device

Ib ≤ In ie. design current must not exceed current setting of device

In ≤ Iz ie. current setting of device must not exceed the lowest conductor rating

∴ Ib ≤ In ≤ Iz   (a)

    I2 ≥ 1.45 × Iz   (b)

For the following devices if the conditions in expression (a) are satisfied then the conditions in expression (b) will also be satisfied:

    HBC fuse to BS 88
    cartridge fuse to BS 1361
    circuit breaker to BS 3781 part 1

Where a semi-enclosed (rewireable) fuse to BS 3036 is used then in order to satisfy expression (b)

In ≤ 0.725 × Iz

2

Module 7 - Formulae (cont'd)

Protection against short circuit currents

For short circuits of duration up to 5 seconds

$$\text{then } t = \frac{k^2 \, S^2}{I^2}$$

Where		
t	=	duration of short circuits in seconds
S	=	csa in mm^2
I	=	effective short circuit current in amperes
k	=	constant value for particular type of cable

CSA of cpc

$$S = \frac{\sqrt{I^2 \, t}}{k}$$

Module 8 - Symbols

$$mV/A/m = \text{millivolt per ampere per metre}$$

$$I_t = \text{Tabulated current-carrying capacity of cable}$$

Module 8 - Formulae

Thermal Insulation

Derating Factors

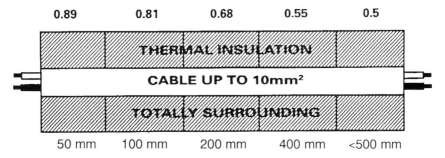

Length in Insulation

$$I_t \geq \frac{I_n}{C_a \times C_g \times C_i} \quad \text{amperes}$$

Where		
C_a	=	correction factor for ambient temperature
C_g	=	correction factor for grouping
C_i	=	correction factor if cable is in contact with thermal insulation

When the overcurrent device is a BS 3036 fuse then the tabulated current carrying capacity of the cable will be

$$I_t \geq \frac{I_n}{C_a \times C_g \times C_i \times 0.725} \quad \text{amperes}$$

0.725 factor is not applicable to MI installation

Module 8 - Formulae (cont'd)

Voltage drop

Must not exceed 4% of the nominal voltage between the origin of the installation and the current using equipment

$$\text{Maximum voltage drop} = \text{voltage} \times \frac{4}{100}$$

$$\text{Actual voltage drop} = \frac{\text{mV/A/m} \times \text{Ib} \times \text{length}}{1000}$$

Module 9 - Formulae

Space factor

$$\text{Cross-sectional area of a cable} = \frac{\pi \times d^2}{4} \quad \text{where d is the overall diameter of the cable}$$

$$A \times \frac{45 \text{ mm}^2}{100} \geq \text{total csa of cables}$$

$$A \geq \frac{100}{45} \times \text{total csa of cables}$$

Where A = the csa of the trunking

Module 10 - Formulae

Cross-sectional areas of protective conductors

$$S = \frac{\sqrt{I^2 t}}{k} \quad \text{mm}^2$$

Where

S is the cross-sectional area in mm^2

I = value of maximum fault current in amperes

t = operating time of the device in seconds

k = factor for specific protective conductors

1

Plan and Style of Regulations

The 16th edition is based on the international regulations produced by the International Electrotechnical Commission (I.E.C.). It is the aim of the I.E.C. to eventually have a common set of wiring regulations.

A large number of the new Regulations are based on I.E.C. Rules already published. For some topics the I.E.C work is still in progress, so the Regulations contain regulations from the previous edition. The 16th edition recognises British Standards, harmonised standards and foreign national standards based on an I.E.C standard.

✳ Format

The Regulations are divided into 7 parts with 6 Appendices. This format is illustrated overleaf.

✳ Numbering

Each part of the Regulations is numbered consecutively, being identified by the first number of each group of digits. The parts are divided into chapters, identified by the second digit and each chapter is split into sections, identified by the third digit. After the first group of three digits, the digits separated by hyphens identify the regulation itself.

Example 1

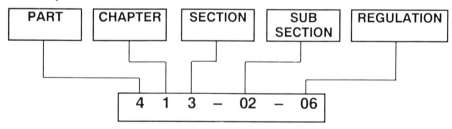

Example 2

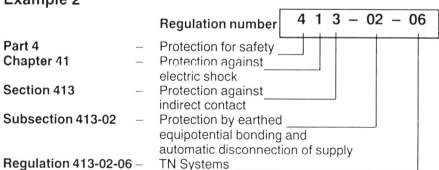

Part 4	–	Protection for safety
Chapter 41	–	Protection against electric shock
Section 413	–	Protection against indirect contact
Subsection 413-02	–	Protection by earthed equipotential bonding and automatic disconnection of supply
Regulation 413-02-06	–	TN Systems

Index

At the beginning of each **PART** of the Regulations, Chapters and Sections are listed; before each **CHAPTER** the Section and their Regulations are noted.

In their new form, the Regulations offer the designer the opportunity of complying with the Regulations by selecting, from various methods described, the one most suited to each particular installation from both economic and technical points of view.

As a result, the detail applying to a particular set of circumstances is sometimes spread across a number of widely separated regulations. None of these individual regulations can be applied in isolation; the overall situation can only be determined by taking into consideration all of the applicable regulations.

Cross referencing related sections may be simplified:

- through the use of the index

   and

- by reference to the diagram on the following page

✳ Relevance to Statutory Regulations and British Standards

The IEE Wiring Regulations have the status of a British Standard. The full title is *British Standard 7671: 1992 Requirements for Electrical Installation (The IEE Wiring Regulations).* The following statutory regulations recognise the IEE Wiring Regulations as a code of good practice.

- Health and Safety at Work Act 1974

- Electricity Supply Regulations 1988

- Electricity at Work Regulations 1989

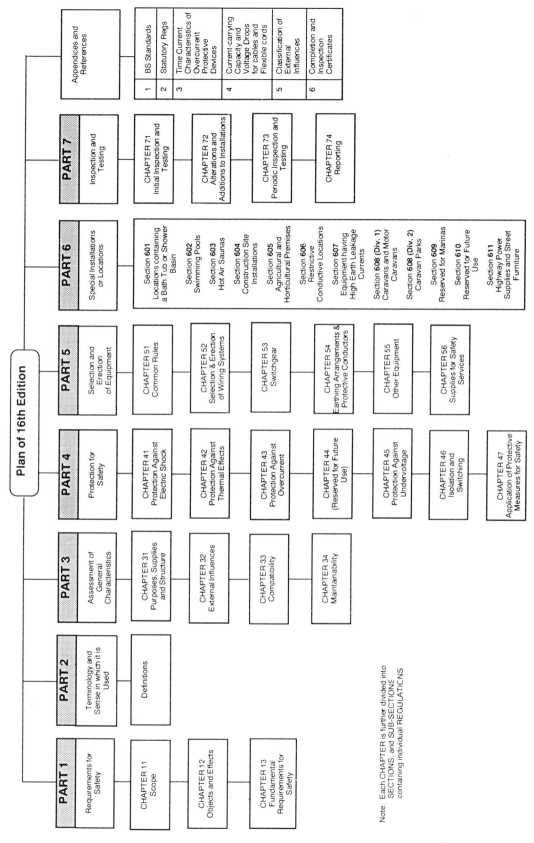

Plan of 16th Edition

PART 1 — Requirements for Safety
- CHAPTER 11 Scope
- CHAPTER 12 Objects and Effects
- CHAPTER 13 Fundamental Requirements for Safety

PART 2 — Terminology and Sense in which it is Used
- Definitions

PART 3 — Assessment of General Characteristics
- CHAPTER 31 Purposes, Supplies and Structure
- CHAPTER 32 External Influences
- CHAPTER 33 Compatibility
- CHAPTER 34 Maintainability

PART 4 — Protection for Safety
- CHAPTER 41 Protection Against Electric Shock
- CHAPTER 42 Protection Against Thermal Effects
- CHAPTER 43 Protection Against Overcurrent
- CHAPTER 44 (Reserved for Future Use)
- CHAPTER 45 Protection Against Undervoltage
- CHAPTER 46 Isolation and Switching
- CHAPTER 47 Application of Protective Measures for Safety

PART 5 — Selection and Erection of Equipment
- CHAPTER 51 Common Rules
- CHAPTER 52 Selection & Erection of Wiring Systems
- CHAPTER 53 Switchgear
- CHAPTER 54 Earthing Arrangements & Protective Conductors
- CHAPTER 55 Other Equipment
- CHAPTER 56 Supplies for Safety Services

PART 6 — Special Installations or Locations
- Section 601 Locations containing a Bath Tub or Shower Basin
- Section 602 Swimming Pools
- Section 603 Hot Air Saunas
- Section 604 Construction Site Installations
- Section 605 Agricultural and Horticultural Premises
- Section 606 Restrictive Conductive Locations
- Section 607 Equipment having High Earth Leakage Currents
- Section 608 (Div. 1) Caravans and Motor Caravans
- Section 608 (Div. 2) Caravan Parks
- Section 609 Reserved for Marinas
- Section 610 Reserved for Future Use
- Section 611 Highway Power Supplies and Street Furniture

PART 7 — Inspection and Testing
- CHAPTER 71 Initial Inspection and Testing
- CHAPTER 72 Alterations and Additions to Installations
- CHAPTER 73 Periodic Inspection and Testing
- CHAPTER 74 Reporting

Appendices and References
1. BS Standards
2. Statutory Regs
3. Time Current Characteristics of Overcurrent Protective Devices
4. Current-carrying Capacity and Voltage Drops for cables and Flexible cords
5. Classification of External Influences
6. Completion and Inspection Certificates

Note: Each CHAPTER is further divided into SECTIONS, and SUB-SECTIONS containing individual REGULATIONS

1/3

2

Scope, Object and Fundamental Requirements for Safety (Part 1)

Scope *(Ref. Chapter 11)*

The regulations apply to the

- design

- selection

- erection

 - inspection and testing of

electrical installations and include particular requirements for electrical installation of:

- locations containing a bath or a shower unit

- swimming pools

- locations containing a hot air sauna

- construction sites

- agricultural and horticultural premises

- restrictive conductive locations

- caravans and caravan parks

- highway power supplies and street furniture

In certain installations the requirements of the IEE Wiring Regulations may need to be supplemented by the requirements of the person ordering the work or by those of British Standards, eg:

- emergency lighting - BS 5266

- installations in explosive atmospheres - BS 5345 (previously excluded)

- fire detection and alarm systems in buildings - BS 5839

Exclusions from Scope

The regulations **do not** apply to

Generation and supply systems

Railway traction equipment rolling stock and railway signalling equipment

Motor vehicles (except those to which the regulations concerning caravans are applicable).

Ships

Offshore installations

Aircraft

Mines and quarries

Radio interference suppression equipment, except where it affects the safety of an installation.

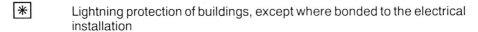 Lightning protection of buildings, except where bonded to the electrical installation

Lift installations covered by BS 5655

The regulations do not apply to the construction of electrical equipment, but only to its selection and application in an installation.

Voltage Ranges

The regulations cover installations with the following operating voltages:

Extra low voltage: 0V to 50V a.c. or 120V ripple free d.c., whether between conductors or to earth

Low voltage: Exceeding extra low voltage to 1000V a.c. or 1500V d.c. between conductors, or 600V a.c or 900V d.c between conductors to earth

Object and Effects, *(Ref. Chapter 12)*

⊠ The regulations are designed to protect persons, property and livestock especially from ELECTRIC SHOCK, FIRE and BURNS, also injury from mechanical movements of electricity operated equipment

They should be cited in their entirety if referred to in any contract.

They are not intended to take the place of detailed specifications, instruct untrained persons, or provide for every circumstance.

The advice of suitably qualified electrical engineers should be obtained for installations which are difficult or of a special character.

Compliance with Chapter 13 of the IEE Regulations will, in general, satisfy the statutory requirements listed in Appendix 2 of the regulations as follows:

Electricity Supply Regulations, 1988

Building Standards (Scotland) Regulations 1990

Electricity at Work Regulations 1989

Cinematograph Regulations 1955 made under the Cinematograph Act 1909, and/or Cinematograph Act, 1952.

Agriculture (Stationary Machinery) Regulations 1959.

Established materials, equipment and methods only are considered, but where alternative materials, equipment and methods affording an equivalent degree of safety are used, these must be the subject of a written specification by a competent body or person who must accept responsibility for its use.

Where statutory control of licensing is exercised the requirements of that authority should also be taken into account, eg. local authority or fire authority.

⊠ When the installation of new material, or inventions not covered by the regulations is necessary, it must be in accordance with the written specification of a competent person or body which ensures that the degree of safety is not less than that obtained by compliance with the regulations.

Details of departures and approvals must bo recorded on the completion certificate.

Fundamental Requirements for Safety *(Ref. Chapter 13)*

These fundamental requirements for safety are identified under the following sub-headings, in very general terms as 20 short regulations. These are later greatly expanded upon in the body of Regulations and Appendices.

Workmanship and Proper Materials *(130 - 01)*

Good workmanship and proper materials shall be used.

General *(130 - 02)*

All equipment to be constructed, installed and maintained, inspected and tested for safety.

All equipment to be suitable for the maximum power demanded when functioning normally.

All conductors to be of sufficient size and capacity for the purposes intended.

All conductors to be either insulated, effectively protected, or placed and safeguarded to prevent damage as far as is reasonably practical.

All joints and connections to be properly constructed as regards conductance, insulation, mechanical strength and protection.

Overcurrent Protective Devices *(130 - 03)*

When necessary, to prevent danger, installations and circuits must be protected against overcurrant by devices which will:

- operate automatically

- be of adequate breaking (and where appropriate) making capacity

- be suitably located and constructed to prevent danger from overheating, arcing, etc., and permit ready restoration of supply without danger.

Precaution Against Leakage and Earth Fault Currents *(130 - 04)*

Precautions are required where metalwork of electrical equipment may become charged with electricity, due to a breakdown in insulation or a fault in equipment, in such a way as to cause danger:

- metalwork is to be earthed in order to discharge electrical energy without danger, or protected by other equally effective means.

Every circuit to be arranged to prevent persistent dangerous earth leakage currents.

Where metalwork of electrical equipment is earthed, the circuits concerned are to be protected against dangerous earth fault currents by overcurrent protective devices or residual current devices, or any other equally effective device.

Where metalwork of electrical equipment is earthed and is accessible at the same time as other exposed metal parts of other services, (eg gas and water), these other services must be effectively connected to the main earthing terminal of the installation if the metalwork of the other services is liable to introduce a potential, generally earth potential.

Protective Devices and Switches *(130 - 05)*

No fuse or circuit breaker, other than a linked circuit breaker, shall be inserted in an earthed neutral conductor.

Single pole switches shall be inserted in the phase conductor only, and any switch so arranged that it also breaks all the related phase conductors.

Isolation and Switching *(130 - 06)*

Isolation and switching arrangements shall be installed so that all voltage can be effectively cut off from every installation or from every circuit and from all equipment as may be necessary to prevent or remove danger. These must be effective and positioned ready for operation.

Every electric motor shall be provided with an efficient means of disconnection, which shall be readily accessible, easily operated and placed so as to prevent danger.

Accessibility of Equipment *(130 - 07)*

Every piece of equipment which requires operation or attention shall be installed so that an adequate and safe means of access and working space is afforded for such operation or attention.

Precautions in Adverse Conditions *(130 - 08)*

All equipment likely to be exposed to corrosive atmospheres and adverse weather or other conditions shall be constructed or protected to prevent danger arising from such exposure.

Where there is risk of fire or explosion all equipment shall be constructed or protected, and other special precautions taken, to prevent danger.

Additions and Alterations to an Installation *(130 - 09)*

No additions or alterations shall be made to an existing installation unless it has been ascertained that the ratings and condition of any existing equipment (including the supply) is adequate for the altered circumstances and any additional load, and that the earthing arrangements are also adequate.

Inspection and Testing *(130 - 10)*

On completion of an installation, or extension or alteration of an installation, appropriate tests and inspection should be made to verify, so far as is reasonably practicable, that the requirements of Regulations 130 - 01 to 09 have been met.

⊛ Persons ordering or requesting an inspection should be informed of the requirements for periodic inspection and testing covered in Chapter 13 of the regulations by the person carrying out the inspection.

3

Assessment of General Characteristics (Part 3)

General

This part of the regulations deals with the need to assess the general characteristics of the energy source or supply and the installation itself.

The following factors are to be assessed and taken into account when choosing the methods of protection for safety (Part 4) and when selecting and erecting equipment (Part 5).

— the purpose for which the electrical installation is to be used, its general structure and supplies (Chapter 31)

— the external influences to which it is exposed (Chapter 32 and Appendix 5)

— the compatibility of its equipment (Chapter 33)

— its maintainability (Chapter 34)

Purposes, Supplies and Structure *(Ref. Chapter 31)*

Maximum Demand and Diversity

The maximum demand of the electrical installation expressed as a current value, must be assessed. Diversity may be taken into account when determining the maximum demand.

Arrangement of Live Conductors and Types of Earthing

The characteristics for number and type of live conductors and earthing arrangements must be assessed and the appropriate methods of protection for safety selected to avoid danger.

Number and Type of Live Conductors *eg. Single phase 2 wire or three phase 4 wire (a.c.)*

The number and types of live conductors, eg. single phase 2 wire or three phase 4 wire (a.c.) for the source of energy and for the circuits to be used in the installation need to be assessed; the energy supplier (eg electricity company) should be consulted where necessary.

Type of Earthing Arrangement

The type of earthing arrangement(s) to be used must also be determined.

The choice of arrangements may be limited by the characteristics of the energy source and any facilities for earthing.

Nature of Supply

The following characteristics should be ascertained for an external supply (and be determined for a private source)

(i) nominal voltage(s)

(ii) the nature of current and frequency

(iii) the prospective short-circuit current at the origin of the installation

(iv) the earth loop impedance of that part of the system external to the installation

(v) suitability for the requirement of the installation, including maximum demand

(vi) type and rating of the overcurrent protective device at the origin of the installation

Where supplies for safety services and standby purposes are required, the characteristics of the source(s) of supply shall be assessed to ensure there is adequate capacity and rating for the operation specified. Consultations must take place with the supplier of energy regarding switching arrangements for safety and standby supplies, especially where the various sources are intended to operate in parallel, or the sources of supply must be prevented from operating in parallel.

Installation Circuit Arrangements

Every installation should be divided into circuits as necessary to:

－ avoid danger in the event of a fault, and

－ facilitate safe operation, inspection, testing and maintenance

⚹ A separate circuit shall be provided for each part of an installation which has to be separately controlled to prevent danger, so that the circuits remain energised in the event of failure of other circuits in the installation; eg. emergency stop circuit controlling the power supply in a workshop, when operated, cuts off the power to the machines but not the lighting which has to be separately maintained to prevent danger.

The number of final circuits required in an installation and the number of points supplied by a final circuit shall be arranged to comply with the requirements for overcurrent protection (Chapter 43), isolation and switching (Chapter 46), and current carrying capacity of conductors (Chapter 52).

Each final circuit must be connected to a separate way in a distribution board, and the wiring of each final circuit should be electrically separate from every other final circuit.

External Influences *(Ref. Chapter 32)*

The chapter dealing with the external influences likely to affect the design and safe operation of the installation is not yet at a stage for adoption as a basis for national regulations. Appendix 5 of the regulations contains some useful information on the subject.

Compatibility *(Ref. Chapter 33)*

An assessment should be made of any characteristics of equipment likely to have harmful effects upon other electrical equipment or other services or likely to impair the supply.

The following characteristics have (for example) been identified:

- transient overvoltage
- rapidly fluctuating loads
- starting currents
- harmonic currents (eg. fluorescent lighting loads and thyristor drives)
- mutual inductance
- d.c. feedback
- high frequency oscillation
- earth leakage currents
- any need for additional connections to earth (eg. for equipment needing a connection with earth independent of the means of earthing of the installation, for avoidance of interference with its operation.

Maintainability *(Ref. Chapter 34)*

Maintainability is also a very important factor to consider when deciding on the design of an installation.

An assessment shall be made of the frequency and quality of maintenance that the installation can reasonably be expected to receive during its intended life. This shall include (where practicable) consultation with the person or body responsible for the operation and maintenance of the installation.

Only then can the regulations be applied so that:

- any periodic inspection, testing, maintenance and repairs likely to be necessary during the intended life can be readily and safely carried out and
- the protective measures for safety remain effective during the intended life and
- the reliability of equipment is appropriate to the intended life.

Electrical Supply Systems

System

An electrical system consisting of a single source of electrical energy and an installation. For certain purposes of these regulations, types of system are identified as follows, depending upon the relationship of the source, and of exposed conductive parts of the installations, to earth:

TN System. A system having one or more points of the source of energy directly earthed, the exposed conductive parts of the installation being connected to that point by protective conductors.

TN-C System. In which neutral and protective functions are combined in a single conductor through the system.

TN-S System. Having separate neutral and protective conductors throughout the system.

TN-C-S System. In which neutral and protective functions are combined in a single conductor in part of the system.

TT System. A system having one point of the source of energy directly earthed, the exposed conductive parts of the installation being connected to earth electrodes electrically independent of the earth electrodes of the source.

IT System. A system having no direct connection between live parts and earth, the exposed conductive parts of the electrical installation being earthed.

Voltage Ranges

Extra Low Voltage - (ELV)

⚹ 0V to 50V a.c. (rms) or 120V d.c., ripple free d.c., whether between conductors or to earth.

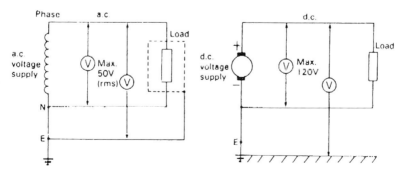

Extra Low Voltage Systems

Low Voltage

Exceeding ELV, but not exceeding 1000V a.c. or 1500V d.c. between conductors or 600V a.c. (rms) or 900V d.c. between conductors and earth.

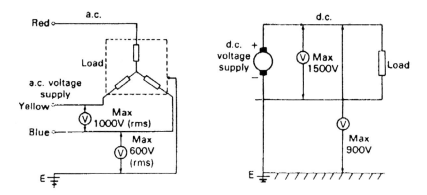

Low Voltage Systems

Classification of Systems

A system consists of an electrical installation connected to a supply. Systems are classified with the following letter designations.

SUPPLY earthing arrangements are indicated by the first letter.

> **T** – one or more points of the supply are directly connected to earth

> **I** – supply system not earthed, or one point earthed through a fault limiting impedance.

INSTALLATION earthing arrangements are indicated by the second letter.

> **T** – exposed conductive parts connected directly to earth.

> **N** – exposed conductive parts connected directly to the earthed point of the source of the electrical supply.

The **EARTHED SUPPLY CONDUCTOR** arrangement is indicated by the third letter.

> **S** – separate neutral and protective conductors.

> **C** – neutral and protective conductors combined in a single conductor.

The types of systems are –

> TN-S   TT   TN-C   TN-C-S   IT

System Earthing Arrangements *(542-1 to 9)*

TN-S Systems

This is likely to be the type of system used where the electricity company's installation is fed from underground cables with metal sheaths and armour. In TN-S systems the consumer's earthing terminal is connected by the supply authority to their protective conductor (ie. the metal sheath and armour of the underground cable network) which provides a continuous path back to the star point of the supply transformer, which is effectively connected to earth.

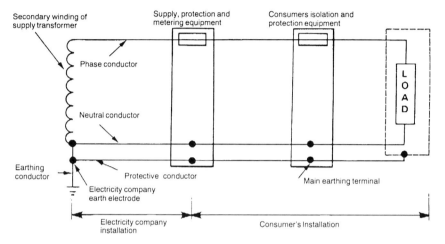

TT Systems

This is likely to be the installation used where the electricity company's installation is fed from overhead cables, where no earth terminal is supplied. With such systems the earth electrode for connecting the circuit protective conductors to earth often has to be provided by the consumer. Effective earth connection is sometimes difficult to obtain and in such cases a residual current device should be installed.

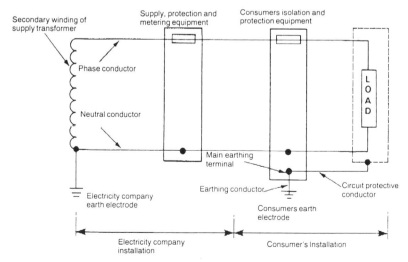

TN-C-S Systems

When the electricity company's installation uses a combined protective and neutral (PEN) conductor, this is known as a TN-C supply system. Where consumer's installations consisting of separate neutral and earth (TN-S) are connected to the TN-C supply system, the combination is called a TN-C-S system. This is the system usually provided to the majority of new installations, referred to as a PME system by the electricity company.

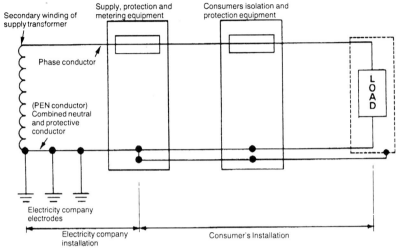

TN-C Systems

Where a combined neutral and earth conductor (PEN conductor) is used in both the supply system and the consumer's installation this would be referred to as a TN-C system.

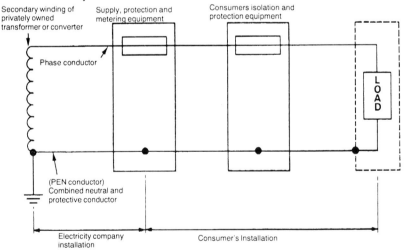

When the consumer's installation comprises earth concentric wiring or another system of PEN conductors, this type of system is usually limited to situations where the installation is supplied by a privately-owned transformer or converter, where there is no metallic connection between this and the public supply system, or where the supply is obtained from a private generating plant.

IT Systems

Where the supply system has either no earth or is deliberately earthed through a high impedance, this is known as an IT system.

With this type of system there is no shock or fire risk involved when an earth fault occurs. The protection is afforded by means of devices which monitor the insulation and give an audible or visual signal or disconnect the supply when a fault occurs.

An IT system must not be used for public supplies. Thus IT systems are generally limited to installations which usually involve a continuous process where the disconnection could result in a hazard and where the installation is not connected directly to the public supply system.

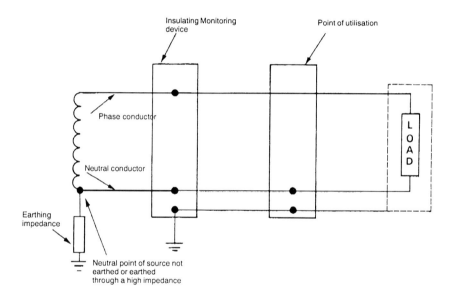

General

In the United Kingdom electricity companies have to comply with the Electricity Supply Regulations 1988.

Full discussions with the relevant electricity company are essential when planning or installing a 'customer's' installation in order to obtain the specifications for any special requirements, eg. size of earthing conductor, earth loop impendance and prospective fault levels for the electrical supply.

✳ S.E.L.V.

An extra-low voltage system which is electrically separated from earth and from other systems in such a way that a single fault cannot give rise to the risk of electric shock.

S.E.L.V. Supplies

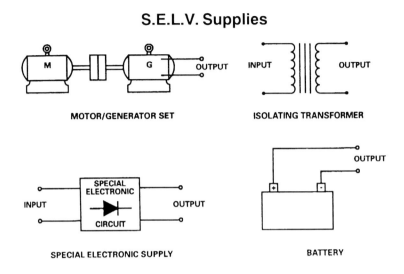

Functional Extra-Low Voltage

Any extra-low voltage system in which not all of the protective measures required for SELV have been applied.

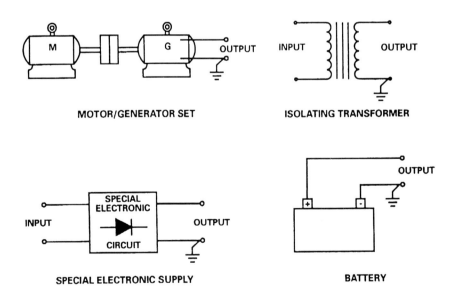

5

Accessories, Diversity and Conventional Circuit Arrangements

Accessories

Plugs and Socket Outlets *(553 - 01)*

Plugs and socket outlets are recognised as being suitable for electrical installations by the IEE Regulations for low voltage circuits (reference should be made to Table 55A of the IEE Regulations).

Plugs and socket outlets for low voltage circuits; types, ratings and relevant British Standards are shown below.

 13 amps
BS 1363 (Fuses to
BS 1362)

 5, 15, 30 amps
BS 196

 2, 5, 15, 30 amps
BS 546 (Fuses, if any,
to BS 646)

 16, 32, 63, 125 amps
BS 4343

These plug and socket outlets are designed so that it is not possible to engage any pin of the plug into a live contact of a socket outlet whilst any other pin of the plug is exposed (not a requirement for SELV circuits) and the plugs are not capable of being inserted into sockets of systems other than their own.

With the exception of SELV or special circuits having characteristics where danger may arise, all socket outlets must be of the non-reversible type, with a point for the connection of a protective conductor.

Plugs and socket outlets other than those shown above may be used on single phase a.c. or 2 wire d.c. circuits operating at voltages not exceeding 250V for the connection of:

Electric clocks — use clock connector unit incorporating a BS 646 or 1362 not exceeding 3 amperes.

[*] Electric shaver socket outlets — use BS 4573

Electric shaver — use BS 3535 shaver unit for use in socket outlets bathrooms.

At construction sites (but not necessarily in site offices, toilets, etc.) only plugs, sockets and couplers to BS 4343 must be used.

Where socket outlets are mounted vertically they should be fixed to a height above floor level or working surface so that the plug and associated flexible cord is not subjected to mechanical damage during insertion, use or withdrawal of the plug.

[*] Cable Couplers *(533 - 02)*

Cable couplers may be used in conjunction with the following types of plug and sockets: (Not to be used on SELV circuits)

BS 196   BS 4343
BS 4491   BS 6991

Cable couplers should be connected so that the plug is on the load side of the installation.

Lampholders *(553 - 03)*

Lampholders must not be connected to any circuit where the value of the overcurrent protective device exceeds those given below (reference should be made to IEE Regulations, Table 55B)

[*] Maximum Rating of Overcurrent Protection Devices of Circuits

		Max. rating
Small Bayonet Cap	B15	6
Bayonet Cap	B22	16
Small Edison Screw	E14	6
Edison Screw	E27	16
Giant Edison Screw	E40	16

This requirement does not apply where the lampholders and their wiring are enclosed in earthed metal or insulating material (ignitability characteristic P BS 476 Part 5) or where separate overcurrent protection is provided.

Lampholders for filament lamps must not be used on circuits operating at voltages in excess of 250 volts.

[✳] For circuits of TN and TT systems the outer contact of Edison type screw lampholders (or single centre bayonet lampholders) must be connected to the neutral conductor. The same requirement applies to track mounted lighting.

[✳] Lighting Points *(553 - 04)*

For each fixed lighting point one of the following must be used.

- ceiling rose to BS 67

- luminaire supporting coupler to BS 6972 or BS 7001

- batten lampholder to BS 5042 or BS 6776

- luminaire designed to be connected directly to the circuit wiring

[✳] When luminaire supporting couplers (LSC) are required they must not be used for the connection of any other equipment; they are designed specifically for mechanical support and the electrical connection of luminaire. If the LSC has a protective conductor contact it cannot be used on a SELV system.

[✳]

Lighting accessories or luminairies must be controlled by a switch or switches to BS 3676 or BS 5518 and must be suitable, where necessary, for the control or discharge lighting circuits.

NOTE: As a rule, for discharge lighting circuits where exact information is not available, the demand in volt-amperes is taken as the rated lamp watts multiplied by not less than 1.8 (reference should be made to Table 1A of the IEE On Site Guide).

Ceiling roses must not be installed in any circuit operating at voltages in excess of 250 volts and must not be used for the attachment of more than one flexible cord, unless specially designed for multiple pendants.

For further details of plugs, socket outlets and general accessories for low voltage circuits conforming to British Standards refer to Technical Data Sheet 5A.

Maximum Demand and Diversity

Diversity

Consider a domestic installation. It is extremely unlikely that all appliances and equipment will be in full use at any one time; for example, in normal circumstances a householder would be unlikely to switch on all the appliances - kettle, fires, water heaters, iron, toaster and cooker - at the same time, and it would be uneconomical to provide cables and switchgear of a capacity for the maximum possible load; the loads they will carry are likely to be less than the maximum. It is this factor which is referred to as 'diversity'. By making allowances for diversity the size and cost of conductors, protective devices and switchgear can be reduced.

To calculate the diversity factor $\left[\dfrac{\text{minimum actual load}}{\text{installed load}}\right]$ for every type of electrical installation, specialist knowledge and experience is required.

⟦✳⟧ The Appendices to 16th Edition of the IEE Regulations do not provide information on this subject; the following information is therefore based on that given in Appendix 1 of the IEE On-Site Guide to the 16th Edition Wiring Regulations.

The common methods of obtaining the current of a circuit is to add together the current demand of all points of utilisation and equipment in a circuit. Typical current demand for points of utilisation and equipment are given in Tables 1A and 1B of the IEE On Site Guide.

Example - Household Cooking Appliances

Application of Diversity Factor

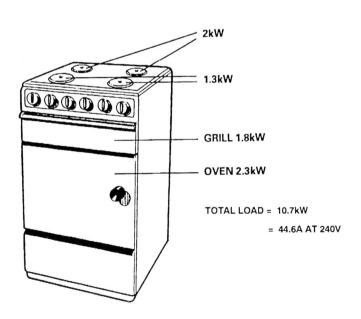

2kW

1.3kW

GRILL 1.8kW

OVEN 2.3kW

TOTAL LOAD = 10.7kW

= 44.6A AT 240V

If we consider, an electric cooker with a maximum loading of 44.6 amperes, as illustrated, the assessed current demand would be as follows:

The first 10 amperes of the total rated current of the cooker, *plus* 30% of the remainder of the total rated current of the cooker, *plus* 5 amperes if a socket outlet is incorporated in the control unit.

Total current rating	44.6A
First 10 amperes = 10	
30% of remaining 34.6 = 10.38	
Socket outlet = 5	
Assessed current demand	25.38A

Discharge Lighting

Final circuits supplying discharge lighting should be capable of carrying the total steady current of the lamp and associated control gear. Where exact information is not available, provided the power factor of the circuit is not less than 0.85, the current demand of a discharge lamp can be calculated from the wattage of the lamp, multiplied by not less than 1.8.

Therefore steady current of a discharge lamp =

$$\frac{\text{lamp power (watts) x 1.8}}{\text{supply voltage}}$$

Example

A circuit supplies five, 240 volt single phase fluorescent luminaires each rated at 65 watts. The current demand would be:

$$1 = \frac{5 \times 65 \times 1.8}{240} = 2.43A$$

Socket Outlets

For conventional power circuits using BS 1363 socket outlets no diversity allowances should be made since this has already been taken into account.

Methods of Applying Diversity

For the design of an installation one of the following methods may be used.

Method 1

The current demand of a circuit supplying a number of final circuits can be obtained by adding the current demands of all the equipment supplied by each final circuit of the system and applying the allowances for diversity given in Table 1B of the IEE On Site Guide. For a circuit with socket outlets the rated current of the protective device is the current demand of the circuit.

Method 2

The alternative method of assessing the current demand of a circuit supplying a number of final circuits is to calculate the diversified current demand for each circuit and then apply a further allowance for diversity, on the assumption that not all circuits will be in use at the same time. The allowances given in Table 1A of the IEE On Site Guide are not applied to the diversity between circuits. The values used should be chosen by the designer of the installation.

Method 3

The current demand of a circuit determined by a suitably qualified electrical engineer.

Example

Consider a small guest house with 10 bedrooms, 3 bathrooms, lounge, dining room, kitchen and utility room with the following loads connected to 240 volt single phase circuits balanced over a 3 phase supply.

Lighting 3 circuits tungsten lighting. Total 2,860 watts

Power 3 x 30 A ring circuits to 13A socket outlets

Water heating   1 x 7 KW shower
                    2 x 3 KW immersion heater thermostatically controlled

Cooking appliances   1 x 3 KW cooker
                       1 x 10.7 KW cooker

Calculations and Answer on next page

5/6

Calculations and Answer to Example

		Current Demand (Amperes)	Table 1B (Diversity Factor)	Current Demand allowing for Diversity (Amperes)
Lighting	$\frac{2.860}{240}$	11.92	75%	8.94
Power	(i)	30	100%	60
	(ii)	30	50%	
	(iii)	30	50%	
Water Heaters (inst)	$\frac{7,000}{240}$	29.2	100%	29.2
Water Heaters (thermo)	$\frac{6,000}{240}$	25	100%	25
Cookers	(i) $\frac{10,700}{240}$	44.58	100%	44.58
	(ii) $\frac{3,000}{240}$	12.5	80%	10

Total Current Demand (allowing Diversity) 177.72

Load - spread over 3 phases =

$$\frac{177.72}{3}$$

$$= 59.24A$$

$$= 60A \text{ per phase}$$

Conventional Circuit Arrangements

(Reference Appendix 9 IEE On-Site Guide)

Types of final circuits using BS 1363 socket outlets and fused connection units are:-

- Radial circuits
- Ring circuits

Radial Circuits

An unlimited number of socket outlets may be supplied, but the floor area which may be served by the socket outlets is limited to either 20m^2 or 50 m^2 depending on the size and type of the cable used and size of overcurrent protection afforded.

Ring Circuits

The requirements for ring circuits are:-

- An unlimited number of socket outlets may be provided. (Each socket outlet of a twin or multiple socket to be regarded as one socket outlet).

- The floor area served by a single 30A ring circuit must not exceed 100 m^2 in domestic installations

- Consideration must be given to the loading of the circuit especially kitchens which may require a separate circuit

- When more than one ring circuit is installed in the same premises, the socket outlets installed should be reasonably shared amongst the ring circuits so that the assessed load is balanced

Cable sizes and overcurrent protection are given below. Reference should be made to Table 9A of the IEE On Site Guide.

Final circuits formats for BS 1363 socket outlets

Circuit	Minimum Conductor Size	Type and Rating of Overcurrent Device	Maximum Floor Area
Ring	2.5 mm^2	30 Amp/32 Amp any type	100 m^2
Radial	4 mm^2	30 Amp/32 Amp MCB or Cartridge Fuse	50 m^2
Radial	2.5 mm^2	20 Amp any type	20 m^2

Values of conductor size may be reduced for fused spurs.

TYPICAL RING CIRCUIT

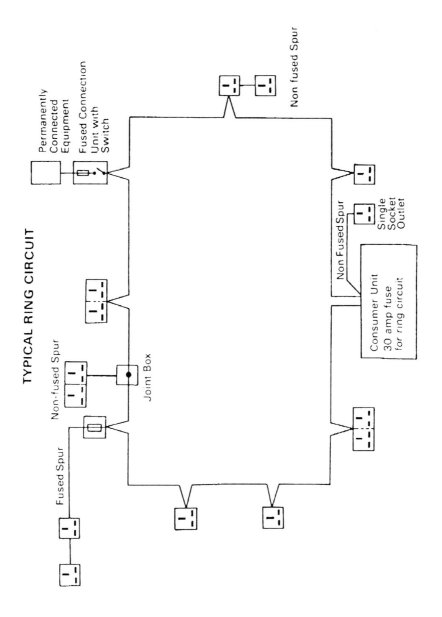

Permanently Connected Equipment

Fused Connection Unit with Switch

Non fused Spur

Non Fused Spur

Single Socket Outlet

Consumer Unit 30 amp fuse for ring circuit

Non-fused Spur

Joint Box

Fused Spur

Spurs

The total number of fused spurs is unlimited, but the number of non-fused spurs must not exceed the total number of socket outlets and any stationary equipment connected directly to the circuit.

Non- Fused Spurs

A non-fused spur may supply only one single or one twin socket outlet or one item of permanently connected equipment.

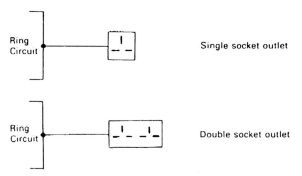

Single socket outlet

Double socket outlet

Permanently Connected Equipment

Permanently connected equipment should be locally protected by a fuse (not exceeding 13A) and be controlled by a switch complying with the Regulations, or be protected by a circuit breaker not exceeding 16A rating.

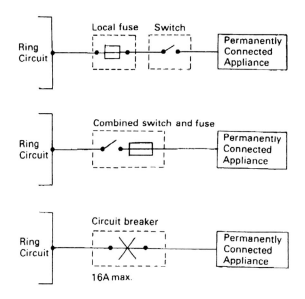

Notes
The cable sizes of non-fused spurs should not be less less than that of the ring circuit
Circuit breaker should be lockable in OFF position

Fused Spurs

A fused spur is connected to a circuit through a fused connection unit. The fuse incorporated should be related to the current carrying capacity of the cable used for the spur, but should not exceed 13A.

When socket outlets are wired from a fused spur the minimum size of conductor is:

1.5 mm^2 for rubber or PVC insulated cables with copper conductors.

2.5 mm^2 for rubber or PVC insulated cables with copper clad aluminium conductors.

1.0 mm^2 for mineral insulated cables with copper conductors.

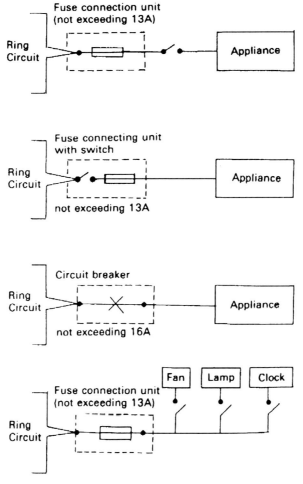

Note
Cable size for fused spur is dependent on the magnitude of the connected load

Method of Connecting Spurs to Circuit

(a) at the terminals of accessories on ring circuit

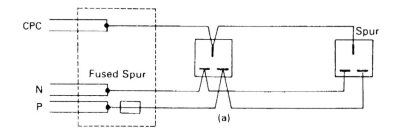

(b) at a joint box

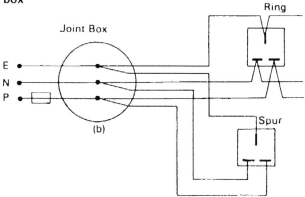

(c) at the origin of the circuit in the distribution board

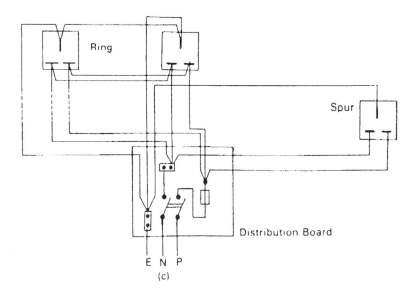

Typical Radial Circuits Using BS 1363 Socket Outlets

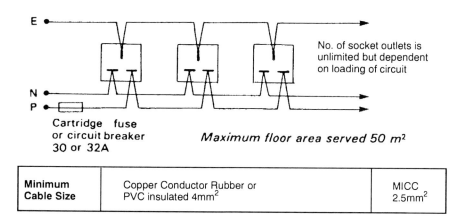

No. of socket outlets is unlimited but dependent on loading of circuit

Cartridge fuse or circuit breaker 30 or 32A

Maximum floor area served 50 m²

Minimum Cable Size	Copper Conductor Rubber or PVC insulated 4mm²	MICC 2.5mm²

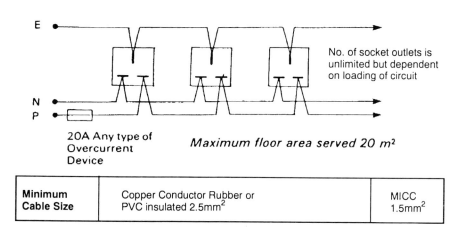

No. of socket outlets is unlimited but dependent on loading of circuit

20A Any type of Overcurrent Device

Maximum floor area served 20 m²

Minimum Cable Size	Copper Conductor Rubber or PVC insulated 2.5mm²	MICC 1.5mm²

Circuits for Immersion Heaters and Space Heating

Where immersion heaters are installed in storage tanks with a capacity in excess of 15 litres, or a fixed comprehensive space heating installation is to be installed, for example, electric fires or storage radiators, separate circuits should be provided for each heater.

Cooker Circuits

A circuit supplying a cooking appliance must include a control switch or cooker control unit which may include a socket outlet. The rating of the circuit should be determined by an assessment of the current demand in accordance with Table 1A of the IEE On Site Guide.

Appliances in a single room *(476-03-04)*

Where two or more appliances are installed in the same room - as is often the case when a fitted kitchen incorporates built-in cooker, hob, microwave oven, fridge and dishwasher - then one single means may be used to control all the appliances, e.g. a series of double pole switches in one panel providing individual control for each appliance

Final Circuits Using BS 196 Socket Outlets

The circuit arrangements for BS 196 socket outlets which are generally installed in industrial installations can be either Radial or Ring circuits.

When assessing the current demand for such circuits (and after allowances for diversity) the maximum current must not exceed 32A which is the largest size of overcurrent device permitted. (Any permanently connected equipment which operates continuously should not have any allowance for diversity).

As the number of socket outlets permitted is unlimited, any spurs must be fused or protected by a circuit breaker of not more than 16A rating.

Types of BS 196 Plugs and Sockets

For circuits with one pole earthed the socket outlet used should be of the type that will accept 2 pole and earth contact plugs, with fusing on the live pole only; the socket should be of the type having raised socket keys and socket keyways recessed at position B.

If the circuit has neither pole earthed (such as a circuit supplied from double wound transformer having the mid point of its secondary winding earthed) the socket outlet must be of the type which will accept 2 pole and earth plugs with double pole fusing. The socket outlets should have raised socket keys, and socket keyway recessed at position P.

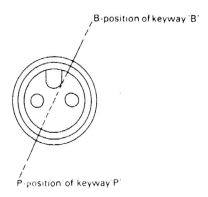

B-position of keyway 'B'

P-position of keyway 'P'

Conductor Size

The conductor size is determined by applying the correction factors in Appendix 9 of the IEE On Site Guide.

Ring circuits - not less than 0.67 times the rating of the protective device.

Radial circuits - not less than the rating of the overcurrent device.

Typical BS 196 Radial and Ring Circuits

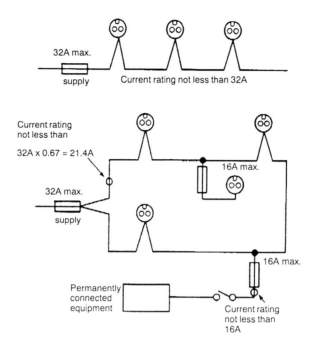

Final Circuits Using 16A BS 4343 Socket Outlets

Plugs and sockets to BS 4343 are available in 16A, 32A, 63A and 125A ratings and are for use in industrial circuits and construction site installations.

The 32A, 63A and 125A rating sockets must be wired on radial circuits each supplying one socket outlet only but the 16A type can be wired in unlimited numbers on radial circuits where the estimated load and diversity permits.

The maximum size of overcurrent device for a BS 4343 multiple 16A socket outlet radial circuit is 20A.

The size of conductor is determined by applying the correction factors from Appendix 4 of the 16th Edition IEE Regulations so that the conductor does not have current carrying capacity less than the rating of the overcurrent protective device.

BS 4343 Socket Outlet Circuits

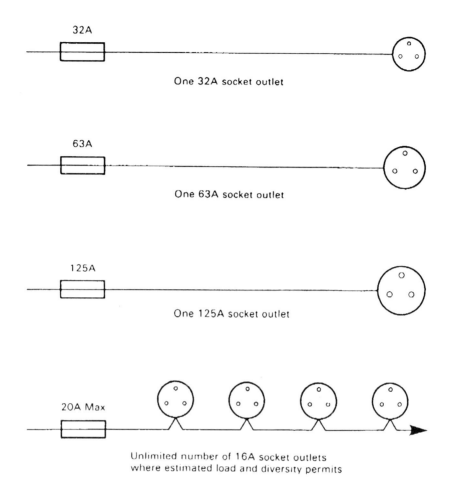

One 32A socket outlet

One 63A socket outlet

One 125A socket outlet

Unlimited number of 16A socket outlets
where estimated load and diversity permits

Extra care must be taken to connect the correct cable conductors to the correct socket outlet terminals.

BS 4343 Socket Outlets

The range of accessories consists of:-

Plugs

Sockets

Cable Coupler

Appliance inlets

Accessories are available for single and three phase supplies with a voltage between phases not exceeding 750V at a rated current of up to 125A.

Discrimination Between Different Voltages

This is achieved in two ways

- By colour codes
- By the positioning of the earth contact in relation to a keyway

Colour code and earth contact relationship to keyway chart are as illustrated below.

Types of BS 4343 Plug and Sockets

Discrimination between accessories of different voltages by position of the earth contact in relation to the keyway.

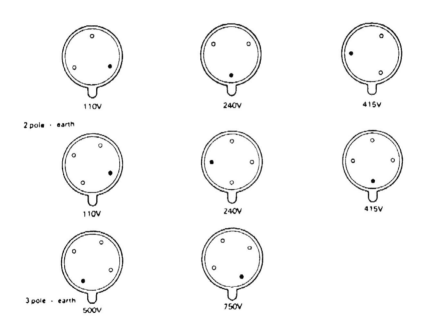

Discrimination between accessories of different voltages by a colour code:

		25V	violet
		50V	white
110V	to	130V	yellow
220V	to	240V	blue
380V	to	415V	red
500V	to	750V	black

Electrical Assessories

13A BS 1363
Socket Outlet

Use with fuses
to BS 1362

Shaver Unit to
BS 4573
(Not for use in
bathrooms).

2, 5, 15 and 30A
Socket Outlets to
BS 546.

Use with fuse to
BS 646 when
necessary.

Shaver Unit to
BS 3535
(For use in
bathrooms)

Incorporates
isolating
transformer

Clock Connector.
Use fuse not
exceeding 3A
to BS 646 or 1362.

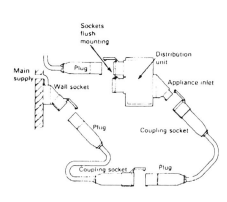

BS 4343 Plugs, Socket Outlets,
Cable Couplers and Inlets

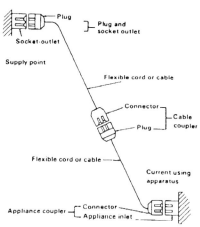

BS 196 Plug Socket and Couplers

Colour coding of fuse links

BS 646

Rating (amps)	Colour code
1	Green
2	Yellow
3	Black
5	Red

BS 1362

3	Red
13	Brown

All other ratings black

6

Protection Against Electric Shock

Introduction

The regulations require that measures shall be taken to protect persons or livestock against both **direct contact** (electric shock in normal service, eg. touching a live conductor), and **indirect contact** (electric shock in case of a fault).

The protection may be afforded by separate measures for both types of contact as stipulated in Section 411 or by a combination of appropriate measures specified in Section 412 for protection against direct contact and Section 413 for protection against indirect contact.

Throughout this chapter, voltage levels are stated. It may be necessary to reduce the voltage level when livestock or reduced body resistance is involved (eg. worktime in damp or wet environments).

※ Protection Against Both Direct and Indirect Contact *(Section 411)*

Many industrial and agricultural installations present situations where the risk of shock to the users of electricity is high, eg. garages, workshops, slaughter houses, power station boiler houses, farms.

Because of the requirements of the safety sources specified, combined protection against both direct and indirect contact is limited to:

- Protection by SELV
- Protection by limitation of discharge of energy

※ Protection by S.E.L.V. *(411 - 02)*

In order to achieve protection by SELV the following requirements must be satisfied.

- Nominal voltage of the circuit shall not exceed extra-low voltage
- The supply source must be one of the following;
  - A safety isolating transformer complying with BS 3535 having no connections between the output winding and the body of the transformer or the protective earthing circuit if any.

    A motor generator with windings providing electrical separation equivalent to a safety isolating transformer complying with BS 3535
  - An electrochemical source (eg. battery)
  - An engine-driven generator
  - Certain electronic devices complying with appropriate standards which ensure that in the event of an internal fault the output voltage cannot exceed the maximum value for extra low voltage 50V a.c or 120V ripple free d.c.

    Refer to Regulation 411-02-09 for details of ripple free d.c.

S.E.L.V. Supplies

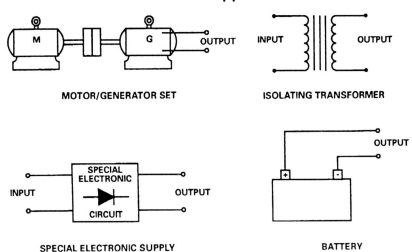

MOTOR/GENERATOR SET

ISOLATING TRANSFORMER

SPECIAL ELECTRONIC SUPPLY

BATTERY

[*] Any system supplied from a system at a higher voltage which does not provide electrical separation such as autotransformers, potentiometers and certain semiconductor devices is not classed as a SELV system.

SELV electrical installations should be installed so that circuits connected to the SELV system are physically separate from any other circuit and have no interconnection with any other electrical system, including earth. Where physical separation of SELV circuits is impracticable, one of the following methods of installation can be used for SELV circuits.

- Conductors insulated to the highest voltage in the installation

- Circuit cable should be of the non-metallic sheathed type.

- Conductors of circuits operating at different voltages must be separated from those at SELV by earthed metal screens (such as found in some trunking systems) or by an earthed metallic sheath (such as MICC or SWA cables).

- A multicore cable may be used to supply SELV circuits as well as circuits operating at different voltages, provided the SELV conductors are insulated for the highest voltage present.

[✻] Under no circumstances should an exposed conductive part of a SELV system be connected to any of the following:

- Earth

- Exposed conductive parts of other systems

- Protective conductors of any system

- An extraneous conductive part, except where electrical equipment is inherently required to be connected to that part, in which case measures must be incorporated to prevent parts exceeding the maximum value for extra low voltage.

Plugs and sockets for SELV systems **must not** be capable of use with other voltage systems in the same premises and must not have a protective conductor contact.

If a SELV does not exceed 25V a.c. (60V ripple free d.c.) no measures to prevent direct contact need to be taken. If there is a risk of electric shock because the body resistance is lower than normal (for example where there are wet conditions or in a confined location such as when working in a large metal pipe or boiler which could act as a conductor) then the voltage limit will need to be reduced.

On installations where the nominal voltage exceeds 25V a.c. or 60V ripple free d.c., protection against direct contact must be provided by barriers or enclosures affording protection to IP 2X (BS 5490) or insulation capable of withstanding a test voltage of 500V d.c. for one minute.

IP 2X (BS 5490) provides that any aperture should not be so large as to allow a British Standard 'finger' through to touch live parts; and never more than 12mm wide. Adequate clearance must be maintained between live parts inside the enclosure.

[✻] Ripple free means for sinusoidal ripple voltage a ripple content not exceeding 10% rms of the peak value. If a luminaire is supplied by an LSC that includes a protective contact, it must be connected to a SELV system.

Protection by Limitation of Energy *(411 - 04)*

Electric fence controllers to BS 2632 used on farms are examples of equipment using this method of protection. In the controller the protection measure is satisfied by limiting the magnitude, duration and frequency of output pulses and of the energy and current that can flow during a specific period.

Circuits relying on this protective measure must be separated from other circuits in the same manner as the safeguards employed for SELV circuits.

Protection Against Direct Contact *(412)*

For protection against direct contact, one or more of the following measures shall be used:

- Insulation of live parts

- Barriers or enclosures

- Protected by obstacles

- Placing out of reach

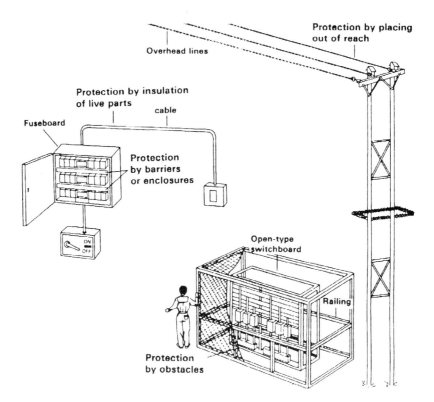

✳ Protection by Insulation of Live Parts *(412 - 02)*

Live parts must be completely covered with insulation which can only be removed by destruction and which is capable of withstanding all stresses in service. These stresses are electrical, mechanical, thermal and chemical.

Insulated cables give protection by insulation of live parts. If insulation is applied during the erection of the installation (such as a switch panel where shrink-on type of insulation is applied to bare copper conductors) the quality of the insulation should be confirmed by tests similar to those which ensure the quality of the insulation of factory built equipment, e.g. flash tested. Paints, varnishes, lacquers and similar products without additional insulation do not comply with the regulations.

Protection by Barriers or Enclosures *(412 - 03)*

For this form of protection live parts must be situated inside enclosures or behind barriers. The barriers and enclosures must be firmly secured in place, have sufficient stability and durability for the known conditions of normal service, and provide protection to IP 2X.

For the top surfaces of barriers and enclosures which are readily accessible the degree of protection is IP 4X.

Alternative requirements for the removal of barriers or the opening of enclosures are stipulated, e.g. by the use of a key or tool such as a screwdriver.

⟨✳⟩ The degree of protection for various situations is stated. Accessories exempt from these regulations are ceiling roses, lampholders, cord operated ceiling switches complying with one of the British Standards; BS 67, BS 3676, BS 5042, BS 6776. Enclosures in common use are conduit, trunking, metal or insulated boxes which have to be opened with a key or tool.

The regulations indicate that one or more measures may be used. Equipment and installations will frequently include several measures as shown in the illustration.

Protection by Obstacles *(412 - 04)*

These regulations require the placing of obstacles which are secured so as to prevent unintentional removal and will prevent accidental bodily approach or unintentional contact with live parts when operating equipment live in normal use. These obstacles (e.g. handrail) may be removed without the use of a key or tool.

It is important to note that this measure can be used in areas accessible only to skilled persons or instructed persons under direct supervision.

Protection by Placing Out of Reach *(412 - 05)*

These regulations apply to overhead lines for distribution between buildings and structures and to bare live parts other than overhead lines.

The out of reach limits set out in the regulations must be increased where bulky or long conducting objects are normally handled, e.g. aluminium ladders or scaffold poles.

It is important to note again that this measure should be used in areas accessible only to skilled persons and instructed persons.

Bare live parts of any installation (not only overhead lines) must not be within arms reach.

Where bare live parts other than overhead lines are out of arms reach but are accessible, they must be installed so that they are not within 2.5 m. of any exposed conductive part, extraneous conductive part or the bare live parts of other circuits.

Supplementary Protection by Residual Current Devices

Residual current devices to BS 4293 with an operating current not exceeding 30mA can be used to reduce the risk of electric shock.

An r.c.d. must not be used as the sole means of protection against direct contact.

Protection Against Indirect Contact *(413)*

The most common method of protection against electric shock from indirect contact is that known as 'earthed equipotential bonding and automatic disconnection of the supply'. The system relies on the fact that all metalwork of the system is connected to earth, and when an earth fault occurs the fault current is sufficient to operate the circuit protective device, e.g. fuse or miniature circuit breaker, in a time not exceeding that given in Regulation 413-02.

When a phase to earth fault of negligible impedance occurs, the potential of the metalwork is raised to a level which is likely to be dangerous.

As shown in the diagram, the phase and protective conductors form a potential divider. As the protective conductor is often of a higher resistance than the phase or neutral conductor due to its reduced c.s.a., a higher proportion of the supply voltage will be developed across it under fault conditions.

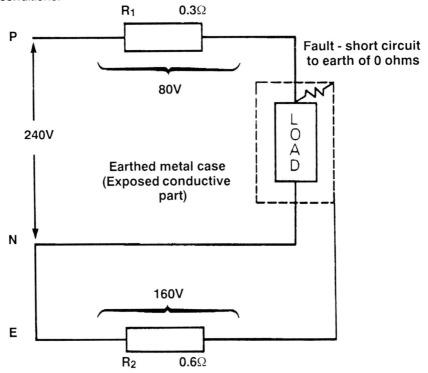

The key requirement of this protective measure is the RAPID disconnection of the supply in the event of an earth fault, this being effected by either an overcurrent protective device or a residual current circuit breaker. Regulation 413-02-04 stipulates the requirements for reducing danger.

The regulations recognise that when an earth fault occurs in an installation the degree of risk is greater with portable equipment held in the hand, such as a portable drill or sander which is gripped continuously, than with fixed equipment. Other factors to be considered are that flexible cables and cords which are used with portable equipment supplied from socket outlets are subjected to a degree of wear and tear and there is no control over the lengths used.

Where these disconnection times are to be achieved by means of overcurrent protective devices, it is necessary to relate the earth fault loop impedance to the type and rating of protective device used to protect the circuit. In order to determine the maximum earth fault loop impedance for a particular circuit it is necessary to refer to the time/current characteristic for that particular device.

Account must be taken of the effect of temperature rise on the resistance values of every circuit conductor during the period it takes to clear a fault.

The requirements of Regulation 413-02-08 are considered to be satisfied when the characteristic of each protective device and the earth fault loop impedance of each circuit protected are such that automatic disconnection of the supply under fault conditions occurs within a specified time. The requirement is met when

$$Zs \leq \frac{Uo}{Ia}$$

where

Zs is the earth fault loop impedance

Ia is the current causing the automatic operation of the protective device within the time given in Table 41A

Uo is the nominal voltage to earth

For 240 volt systems the maximum disconnection time is 0.4 seconds

The maximum disconnection times given in Table 41A apply to circuits supplying socket outlets and any other final circuits which supply portable equipment moved manually, or hand-held class 1 equipment.

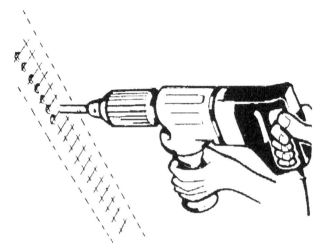

The above requirement does not apply to a final circuit supplying stationary equipment fed via a plug and socket outlet, where precautions prevent the use of the socket outlet for supplying hand-held equipment; nor to the reduced low voltage circuits described in Regulations 471-15 (eg. BS 4363 distribution equipment as illustrated).

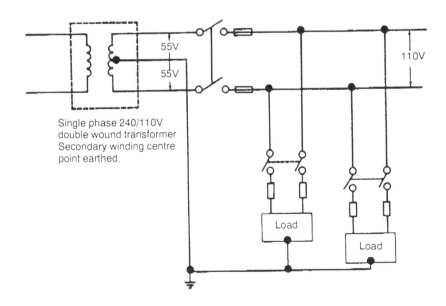

Single phase 240/110V
double wound transformer
Secondary winding centre
point earthed.

Reg: 413-02-10

When fuses are used to meet the requirements of Regulation 413-02-09, the maximum values of loop impedance (Zs) for disconnection times of 0.4 seconds are given in IEE Regulations, Table 41B1, and when circuit breakers are used, IEE Regulations, Table 41B2 gives the maximum loop impendance values.

Reg: 413-02-13

A disconnection time not exceeding 5 seconds is allowed for distribution circuits supplying only stationary equipment and for final circuits where Regulation 413-02-09 does not apply. Tables 41D for fuses and 41B2 for MCBs give maximum values of Zs.

Note: *Part 6 of the Regulations 'Special Installations or Locations - Particular Requirements' contains sections covering specific installations (e.g. on construction sites and at swimming pools) where the risk of shock is considered greater. A modified version of Table 41A is provided; this reduces the value of the disconnection time from 0.4 to 0.2 seconds. Modified Tables 41B1 and 41B2 are also included in these sections.*

The earth fault loop impedance (Zs) is made up of the impedance of the consumer's phase and protective conductors R_1 and R_2 respectively, and the impedance external to the installation Z_E (i.e. impedance of the supply). As the value of Z_E will be obtained from the electricity company for the initial assessment of the installation, the maximum impedance allowed for the phase and protective conductors can be determined from:

$$R_1 + R_2 = Z_S - Z_E$$

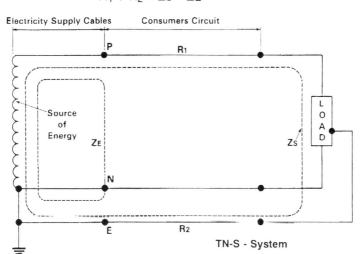

TN-S - System

Further references and details of the external impedance value will be explained in Module 7 - Protection Against Overcurrent.

This limit on the value of $R_1 + R_2$ must now effectively limit the length of run for the circuit cable.

Note: *The 16th Edition does not contain an appendix giving the resistance values per metre for cables. Reference should be made to the manufacturer's literature, or Tables 6A and 6B of the IEE On-Site Guide.*

Use of Values of Cable Resistance Tables

Example 1

A 240V, 5A circuit supplies a number of lighting points in an installation (which includes a bathroom). The circuit is protected by a 5A BS 3036 fuse and is wired in 1.0 mm^2 PVC sheathed cable in a 24 metre length of run. If the value of Z_E is 1.3 ohm, determine if the circuit complies with Regulation 413-02-08. Assume an ambient temperature of 20°C.

Step 1

Find $(R_1 + R_2)$ milli-ohms/metre value from Table 6A of the IEE On Site Guide for 1.0 mm^2 cable

$= 36.2$ milli-ohms/metre for 20°C

Step 2

Apply multiplier for PVC insulation (1.38) from Table 6B of the IEE On Site Guide

Step 3

Find resistance of $(R_1 + R_2)$ for cable run of 24 m

$= 36.2 \times 1.38 \times 24$ m

$= 1198$ milli-ohm (1.2 Ω)

Step 4

Use formula $Z_S = Z_E + (R_1 + R_2)$

$\therefore Z_S = 1.3 + 1.2 = 2.5\ \Omega$

Step 5

Find max. value of loop impedance (Z_S) for circuits supplying fixed equipment from Table 41B1(c) or Table 2A of the IEE On-Site Guide

$= 10\Omega$

Circuit does comply with Regulation 413-02-08

Example 2

A 240V, 20A socket outlet circuit is to be protected by a BS 88 Part 2 fuse and is wired in 2.5 mm^2 PVC sheathed cable with a 1.0 mm^2 protective conductor. If the value of Z_E is 1.0 ohm determine the maximum length of run for compliance with Regulations 413-02-08. Assume ambient temperature 20°C.

Step 1

Find (Z_S) max from Table 41B1(a) or Table 2B of the IEE On-Site Guide

$= 1.85\Omega$

Step 2

Find max. value of $(R_1 + R_2)$

$= Z_S - Z_E$

$= 1.85 - 1.0$

$= 0.85 \ \Omega$

Step 3

Find $(R_1 + R_2)$ milli-ohms/metre value from Table 6A of the IEE On Site Guide

$= 25.51$ milli-ohms/metre

Step 4

Apply multiplier (1.38) from Table 6B of the IEE On Site Guide

$= \dfrac{25.51 \times 1.38}{1000}$

$= 0.035 \ \Omega/m$

Step 5

Find max. length of cable run for compliance

$= Z_S - Z_E = \text{max} (R_1 + R_2) - (R_1 + R_2) \ \Omega/m$

$= \dfrac{0.85}{0.035}$

$= 24.28 \ m$

The Alternative Method for Socket Outlet Circuits
(413-02-12)

Another method of determining the earth fault loop impedances for socket outlet final circuits, which will satisfy the requirements of Regulations 413-02-04 has become known as the 'alternative method' and permits the disconnection time to be increased from 0.4 sec to 5 sec., this being the same as for a final circuit supplying fixed equipment indoors. This alternative method is applicable only where the exposed conductive parts of the equipment concerned and any extraneous conductive parts are situated within the zone created by the main equipotential bonding.

Before examining the use of the alternative method, it is necessary to look at the 'touch voltage curve' which was contained in an IEC document. The curve (see fig. on opposite page) shows values of time for which values of touch voltage can be allowed to persist without danger to a person. The curve shows that automatic disconnection of the supply is not required when the voltage is 50V or less.

✳ The alternative method may be useful when the maximum loop impedances of Table 41B1 or Tables 2A(i), 2B(i) and 2D(i) of the IEE On-Site Guide cannot be achieved. It allows the disconnection time to be extended to 5 sec. provided the impedance values given in Table 41C are not exceeded.

Table 41C gives maximum impedance values for protective conductors. These maximum impedance values are based on the principle that the current flowing through the circuit when 50V exists across it will cause disconnection within 5 sec. for any type and rating of overcurrent protective device.

Examples using the tables:

(i) For a 20A cartridge fuse to BS 1361 protecting a socket outlet circuit wired in PVC cable in steel conduit use Table 41C(b), IEE Regulations

then maximum impedance ≤ 0.61 Ω

(ii) For a 30A semi-enclosed (rewireable) fuse to BS 3036 protecting a socket outlet circuit wired in PVC twin cable with c.p.c. use Table 41C(c), IEE Regulations

then maximum impedance ≤ 0.58 Ω

Touch Voltage Curve

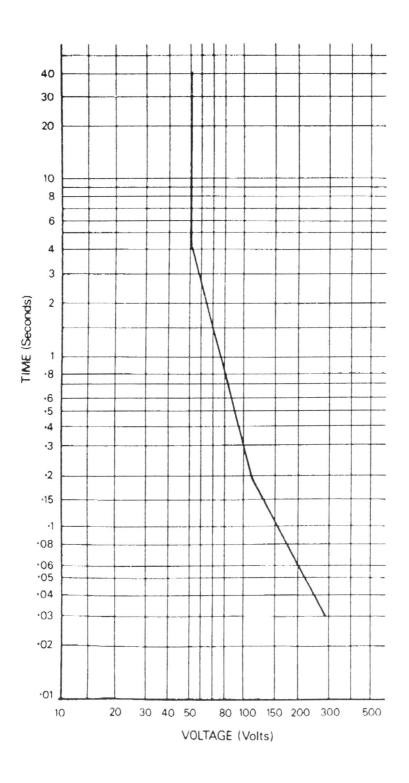

Limiting the impedance of the c.p.c. will ensure that the maximum value appearing between the far end of the c.p.c. and main earth terminal must lie within the safe area of the touch voltage curve. The impedance values ensure that if the disconnection time for an earth fault exceeds 5 sec., the maximum prospective shock voltage will be less than 50V.

Example 1

For a BS 88 Part 2, 25A fuse supplying a circuit using a cable with a separate c.p.c. the maximum impedance value is 0.5 Ω (from Table 41C(a), IEE Regulations)

When 50V appears across the c.p.c. then fault current

$$\text{If} = \frac{50}{0.5} = 100A$$

From the time/current characteristic for a BS 88 Part 2 fuse it can be seen that 100A will cause disconnection in 5 seconds. If the p.d. is greater than 50V the disconnection time will decrease and the 'touch voltage curve' is satisfied. (Time/current characteristics are detailed in Module 7).

Ignoring the effect of the external impedance ZE so that the full voltage Uo appears at the origin of the circuit;

If the phase conductor and the c.p.c. have the same c.s.a. then 120V could appear across the c.p.c.

$$\therefore \text{If} = \frac{120}{0.5} \doteq 240A$$

For the same type of fuse as above the time/current characteristic shows that under these conditions disconnection is achieved in 0.1 sec.

From the 'touch voltage curve' the time is 0.17 secs. Thus the circuit will disconnect safely.

For Equipment to be used outside the Equipotential Zone

A further set of conditions has to be applied for equipment to be used outside the zone in which the circuit feeding that equipment originated (e.g. electric lawn mowers and hedge trimmers supplied from socket outlets rated at 32A or less).

Attention must be given to the requirements of Regulation 471 - 16 which specifies that such equipment must be protected by a residual current device (r.c.d.) with an operating current not exceeding 30 mA.

In domestic premises the location of such an outlet would typically be in the garage. A person using equipment from such an outlet will be in direct contact with the general mass of earth, may come into contact with metal garden furniture and may have reduced body resistance.

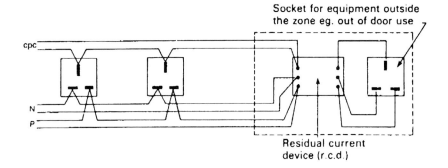

Socket for equipment outside the zone eg. out of door use

cpc

N

P

Residual current device (r.c.d.)

For details of Residual Current Devices see Technical Data Sheet 6A.

Protection Against Indirect Contact

Further methods of providing protection against electric shock from indirect contact are:

- use of Class II equipment

- non-conducting location

- earth free local equipotential bonding

- electrical separation

Where lids or doors of insulating enclosures can be opened without a tool or key, all accessible conductive parts must be placed behind insulating barriers providing protection to IP 2X and must be removable only by use of a tool (eg. protective barriers in distribution boards covering the neutral bar).

Use of Class II Equipment

With Class II equipment (double insulation), the first line of defence for protection against indirect contact is its basic insulation. The second line of defence takes the form of a second layer of supplementary insulation. The outer case of such equipment need not necessarily be insulated.

An exposed metal case of such a Class II appliance is not likely to become live in the event of a fault in the basic insulation and is not therefore an exposed conductive part (as defined). Such items should never be earthed as this would mean that they must comply with the requirements of Class 1 equipment; i.e. all the electrical parts would be continuous with other items of exposed metalwork.

✳ Protection by Non-Conducting Location *(413-04)*

This is a rather special arrangement and has only very limited application. It can only be used in circumstances under effective supervision and where specified by a suitably qualified engineer. This method essentially requires:

- non-conducting floors and walls

- no protective conductors and no earth contact in any socket outlet

- physical separation between exposed conductive parts and extraneous conductive parts (must exceed 2 m)

These situations require strict supervision as they can become dangerous because of the possibility of the introduction of earthed metal into the location in the form of a portable appliance fed by leads connected outside the location. The safety of such a system is in the fact that a potential reached by metalwork in such a situation is unimportant because it is never possible to touch two pieces of metalwork at different potentials at the same time.

✳ Protection by Earth Free Local Equipotential Bonding *(413 - 05)*

Earth free local equipotential bonding can only be used in an area under effective supervision and where specified by a suitably qualified electrical engineer. It relies on the fact that all simultaneously exposed and extraneous conductive parts are connected together by bonding conductors, but not to earth. Inside the area there can be no danger, even if the potential is high, because all internal exposed metalwork is at the same potential.

Protection by Electrical Separation *(413 - 06)*

Electrical separation requires the item to be supplied via a safety isolating transformer complying with BS 3535, the secondary of which is unearthed. It is important to ensure that the separated circuit is not accidentally earthed, because should this happen a protected person would be at risk from voltages which could then appear between simultaneously accessible parts in the event of a second earth fault. All precautions must be taken to ensure as far as possible that earth faults cannot occur.

A common example of this protection method is a bathroom shaver unit feeding an electric shaver.

Other methods include a motor generator with windings providing the safety equivalent to a safety isolating transformer.

Residual Current Devices (RCD)

An r.c.d. is designed to give protection against shock risk and against fire. The basic circuit for a single phase device is as illustrated.

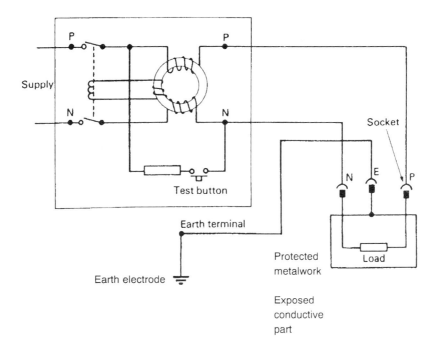

Operation

The current taken by the load is fed through two equal and opposing coils wound on to a common transformer core. When the phase and neutral currents are balanced (as they should be on a healthy circuit) they produce equal and opposing fluxes in the transformer core resulting in no voltage in the trip coil. If more current flows in the phase than in the neutral an out-of-balance flux will be produced which will be detected by the fault detector coil. The fault detector coil opens the DP switch by energising the trip coil.

Test Switch/Button

The test switch tests only that the circuit breaker is functioning correctly and is operating in the correct order of sensitivity, as specified by BS 4293.

Selectivity

Rccb's are completely selective in operation of the circuit they protect and are unaffected by parallel earth paths.

Application

The device must be capable of disconnecting all phase conductors of the circuit. The live conductors of the circuit must be contained within the magnetic field of the transformer of residual current devices. Any protective conductor must be outside this magnetic field, eliminating any possibility of an induced emf in the protective conductor.

If the operation of the device relies upon an auxiliary supply which is external to the device, it must be of a type that will automatically operate if the auxiliary supply fails. Alternatively the device may be provided with a supply which will automatically become available upon failure of the auxiliary supply.

Residual current devices should be installed outside the magnetic fields of other equipment (unless the manufacturer gives instructions to the contrary).

When a device is fitted as protection against indirect contact but separately from overcurrent protection, the device must be capable of withstanding without damage any thermal and mechanical stress which may occur under short circuit conditions on the load side of the device.

If the disconnection times are to be achieved by the use of a residual current device (rcd) in an installation supplied from a TN or TT system, the product of the rated residual operating current (A) and the earth fault loop impedance (Ω) must not exceed 50 Volts.

Note: The use of an rcd is excluded for automatic disconnection when the system is TN-C. In such a case there is no difference between phase and neutral currents because there is no separate path for neutral and earth leakage currents.

7

Protection Against Overcurrent (Chapter 43)

General (431)

This chapter gives the requirements for circuits to be protected against overcurrents and makes a distinction between overload current and fault current.

In considering protection, Part 4 of the Regulations is concerned throughout with the safe protection of persons, livestock and property. A section (435) also deals with the need to co-ordinate the protection against the two forms of overcurrent.

Prospective Fault Current

The prospective short circuit current (I_p) is an expected value of current which can be calculated as follows:

$$I_p = \frac{V}{Z_t + Z_1 + Z_2}$$

where V = source voltage
Z_t = impedance of supply transformer
Z_1 Z_2 = conductor impedances
($Z_2 = Z_n$ for single phase and neutral circuit)

Note: The definition covers the condition of negligible impedance, i.e. there is in effect a 'fault' between the phase conductors, or between phase and neutral conductors in the installation (see diagram below).

The determination and application of the fault current is dealt with later.

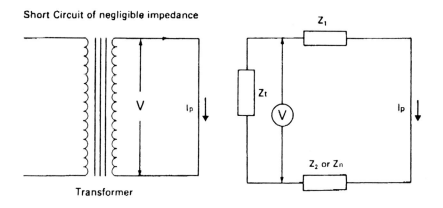

Short Circuit of negligible impedance

Transformer

Protection Against Both Overload and Fault Currents *(432-02)*

To be able to give this protection there is a requirement for a device which must be capable of making and breaking any over-current up to and including the prospective short circuit current (I_p) at the point where the device is installed.

Types of protective devices available:

Fuses

Circuit breakers incorporating overload release

Circuit breakers in conjunction with fuses

For description, construction, operation and application of these devices see Technical Data Sheets 7A, 7B, 7C.

Protection Against Overload Current Only *(432-03)*

The device used must be capable of breaking the circuit before any overload current could cause a rise in temperature which might damage insulation, terminations, joints or surroundings of conductors.

These requirements mean that the circuit protective device is selected to ensure that overheating does not occur to a degree of breakdown of insulation.

It is sensible to design an installation in the most economic manner. Savings can be made if attention is given to the requirements of the Regulations covering conductors and protective devices.

Quantities to be considered are:

I_n = nominal current or current setting of device

I_b = design current of the circuit

I_z = current-carrying capacity of any of the circuit conductors

I_2 = the current which ensures effective operation of the device

Regulation 433-02 requires the effective operating current (I_2) to be not greater than 1.45 times the current carrying capacity of the related conductor. This takes into account the fact that a protective device will require a current larger than its rating to operate it.

The co-ordination of requirements in this case are:

that $I_b \leq I_n$; i.e. design current must not exceed current setting of device

that $I_n \leq I_z$; i.e current setting of device must not exceed the lowest conductor rating

$$\therefore \; I_b \leq I_n \leq I_z \quad \text{(a)}$$

$$\therefore \; I_2 \leq 1.45 \times I_z \quad \text{(b)}$$

when the device is either:

− an HBC fuse to BS 88 Part 2

− an HBC fuse to BS 88 Part 6

− or a cartridge fuse to BS 1361

− or a circuit breaker to BS 3871 Part 1

If the conditions in expression (a) are satisfied then the conditions in (b) will also be satisfied.

When the device is a semi-enclosed fuse to BS 3036, in order to satisfy expression (b) above $I_n \leq 0.725 \, I_z$. The reason for this is due to the possibility of such a fuse having a fusing factor as high as 2, then

✳

fusing current = fusing factor x fuse rating
$\qquad\qquad\quad$ = 2 x fuse rating

then $I_2 = 2 \, I_n$

and to satisfy (b) $2 \, I_n \leq 1.45 \times I_z$

$$\therefore \; I_n \leq \frac{1.45 \times I_z}{2} \leq 0.725 \times I_z$$

In practice then, if a BS 3036 fuse is used as the protective device, this part of Regulation 433-02 may mean that a larger cable than that determined from normal load conditions must be selected from the cable rating tables.

Current carrying capacities of cables and flexible cords (I_z values) are given in Appendix 4 of the IEE Regulations for a range of installation methods, using either copper or aluminium conductors.

Time/Current Characteristics

The disconnection time for overcurrent devices such as fuses and miniature circuit breakers to operate is given by a time/current characteristic curve. Typical time/current characteristics are given in Appendix 3 of the IEE Regulations.

It can be seen that at very high currents the operation of the overcurrent devices takes place in a very short period of time; for small overloads the time taken is longer.

The method of presenting the time/current characteristics is by a graph with the time/current axes scaled logarithmically. This means that each successive graduation represents a ten times change of the previous graduation.

Each graduation of time being labelled

> 0.01s   0.1s   1.0s   10s   100s

and graduations of current are labelled

> 1A   10A   100A   1000A

The method of reading time/current characteristics is illustrated below:

- when the fault current I_f = 30A the time taken to operate = 0.56 seconds

- when the fault current I_f = 180A the time taken to operate = 0.12 seconds

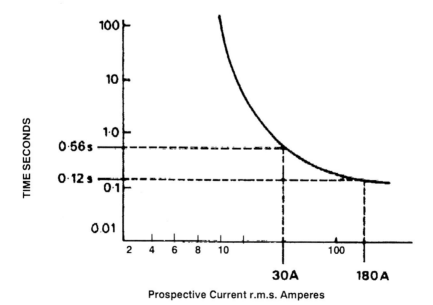

Prospective Current r.m.s. Amperes

✳ The time current characteristics of overcurrent protective devices in Appendix 3 of the IEE Regulations have been simplified. The axis are now easier to read and a table has been included, indicating worked values as shown below:

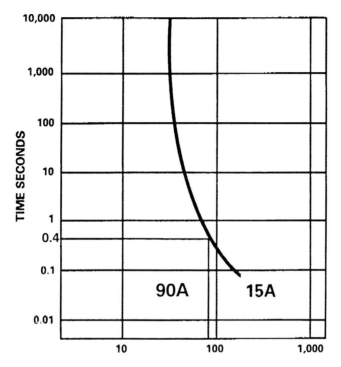

Prospective current in Amperes

TIME/CURRENT CHARACTERISTICS				
FUSE RATING	CURRENT FOR TIME			
	0.1 SEC	0.2 SEC	0.4 SEC	O.5 SEC
5A				
15A			90A	
30A				
60A				

✳ **Overload Protection of Conductors in Parallel** *(433-03)*

Except in the case of a ring final circuit, a single protective device such as a fuse or circuit breaker may be used to protect two cables run in parallel supplying a single piece of equipment, providing the cable conductors are the same type, cross-sectional area, length, installed in the same manner and no other circuits will be supplied from that cable.

Position of Devices *(473-01)*

In the application of overload protective devices, particular attention must be given to the position of a device in a circuit.

A device for overload protection must be placed at a point where reduction occurs in the current-carrying capacity of the conductors (exceptions are referred to below).

Examples of how this reduction may occur are:

- Reduced conductor c.s.a., e.g. a fused spur on a ring circuit

- Installation method changed (e.g. over-head to underground)

- Type of cable changed, e.g. PVC cables in conduit changed to MICC

- Ambient temperature changed (e.g. from a boiler house to normal room conditions)

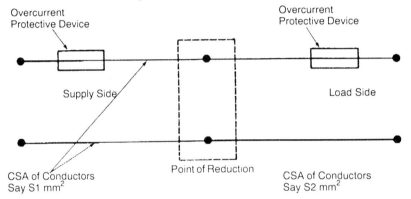

Provided that there are no outlets or spurs along a section the device may be placed along the cable run.

Conditions for omission of devices for overload protection

Devices for protection against overload **need not** be provided:

- On the load side of the point where a reduction occurs in the current-carrying capacity, where the conductor is effectively protected against overload by a protective device placed on the supply side

- Where, because of the characteristics of the load or the supply a conductor is not likely to carry overload current

- In a circuit supplying equipment where unexpected opening of the circuit causes a greater danger than an overload condition

- At the origin of an installation where the supplier provides an overload device and agrees that it affords protection to the part of the installation between the origin and the main distribution point of the installation where further protection is provided

In the case of current transformers an overload protection device must not interrupt the secondary circuit.

Protection Against Fault Current *(434)*

The requirement is for short circuit currents to be interrupted in a sufficiently short time to prevent danger and damage from the thermal and mechanical effects on conductors or connections.

Emphasis is given in this Regulation the speed of interruption (i.e. rapid operation of the device) in addition to the removal of the dangerous condition.

Determination of Prospective Fault Current (I_p) *(434-02 to 434-03)*

The Regulations require that the prospective short circuit current (I_p) at every relevant point of the complete installation must be determined.

This is done either by measurement or calculation. Measurement has limited application because it is concerned with existing supplies. The calculation requires knowledge of the system impedance. The value of I_p obtained is usually at the load terminals of the protective device at the distribution board.

Local electricity companies will quote a maximum value of I_p and this must be recorded since it will form part of the installation design data. The value will also be required on the Inspection Certificate.

The Regulations require that the breaking capacity or fault current capacity of the device shall be not less than the I_p at the point of installation, the exception being that a lower capacity is permitted if another device with the necessary capacity is installed on the supply side.

Overload protection can provide fault current protection if the protective device has a breaking capacity or short circuit capacity of not less than I_p and the conditions expressed in (a) and (b) (of page 7/3) for protection against overload current are satisfied. In these circumstances such a device can be used to give short circuit protection.

The Regulations state that if the above conditions do not apply, *the time to interrupt the fault current* must be determined to ensure that the temperature limit of the conductors is not exceeded. This is unlikely to be required for conductors of c.s.a. less than 10 mm^2 and for short circuits of duration up to 5 sec.

The formula to be used is $t = \dfrac{k^2 s^2}{I^2}$

for a short circuit condition where the "let-through" energy of the device will be I^2t (details given in Technical Data Sheet 7C).

where t = duration of short circuit in seconds
       S = c.s.a. in mm^2
       I = the value of fault current in amperes, expressed for a.c. as the r.m.s. value, due account being taken of the current limiting effect of the current impedance
       k = factor for a particular type of cable (Table 43A, IEE Regulations gives values for common materials). The k value is dependent upon material of conductor, the conductor size, its initial and final temperature and insulation.

In order to determine the time (t) and to show that it comes within the limit of 5 seconds when a short circuit occurs, it must be calculated. If greater accuracy is required refer to BS 7454.

Example

Cable size and type - 25 mm^2 PVC insulated and sheathed with copper conductors.

Value of k = 115

Take the effective fault current as 4 kA = 4000A

Time permitted for the fault current to exist before damage to the cable insulation could occur is:

$$t = \frac{k^2 S^2}{I^2}$$

$$= \frac{115^2 \times 25^2}{4000^2}$$

$$= 0.52 \text{ sec.}$$

Taking a 50A mcb Type 2 as the protective device, check that circuit breaker will clear the short circuit within 0.52 sec. Refer to the time/current characteristic, (Fig. 5, Appendix 3 of the Regulations) and read off the time value when the current is 4000A.

Time obtained = 0.01 sec. so that the mcb will prove satisfactory.

✳ Position of Devices (473)

As for overload protection, there are certain conditions to be satisfied when deciding on the position of fault current protective devices.

The devices may be positioned on the load side of a point where there is a reduction in rating, subject to a restriction of 3 m maximum length and where the risk of fault current fire or other danger is reduced to a minimum.

Switchgear Connections

An example of the use of this relaxation is when connecting items of switchgear, e.g. switch fuse to busbars.

There is no limit stated for the size of the conductor which can be used. The risks mentioned may be reduced by enclosing the conductors in trunking or conduit.

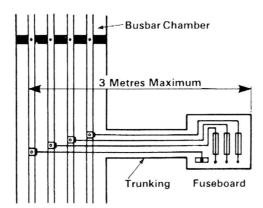

Fault Current Protection of Conductors in Parallel

Except in the case of a ring final circuit, a single protective device such as a fuse or circuit breaker may be used to protect two cables run in parallel supplying a single piece of equipment, providing the cable conductors are the same type, cross-sectional area, length, installed in the same manner and no other circuits will be supplied from that cable. Compliance with Regulation *434-03-03* must be verified by a calculation. Account needs to be taken of the conditions that would occur in the event of a fault not affecting all conductors.

Co-ordination of Protection *(435)*

There is a requirement to co-ordinate overload and fault current protection by assessing the characteristics of the devices used, so that the energy let through by the short circuit protective device does not exceed that which can be withstood without damage by the overload protective device.

This Regulation is particularly applicable to motor control circuits where the overload and short circuit devices may be housed separately.

For details of operational characteristics of devices refer to the Technical Data Sheet 7C.

Limitation of Overcurrent by Characteristics of Supply *(436)*

The conductors of an installation are regarded as protected against both overload and short circuit current in cases where they are fed from a source of supply incapable of supplying a current greater than the current carrying capacity of the conductors.

Overcurrent Protection Devices *(533)*

Replacement of Fuse Links *(533-01)*

If there is a possibility of a person (other than a skilled or instructed person) replacing links used for overload protection, care must be taken to ensure that:

- the fuses are marked with an indication of the type of link to be used

- they are of the type that cannot be inadvertently replaced with a link having the same nominal current rating but a higher fusing factor than that intended

Additional Notes on Fuses *(533-01-04)*

Fuses shall preferably be of the cartridge type but where semi-enclosed (rewireable) fuses are used the fuse element must be fitted in accordance with manufacturer's instructions, e.g. correct c.s.a. and length of element.

In the absence of such instructions the fuse must be fitted with a single element of plain or tinned copper wire of the appropriate diameter. (Refer to Table 53A of the IEE Regulations).

Where fuses have to be removed or replaced whilst the circuit they protect is still energised, the fuses must be so designed so that the operation can be carried out without danger

Operation of Circuit Breakers *(533-01-05)*

Where circuit breakers may be operated other than by skilled or instructed persons, they should be so designed or installed that altering their setting or calibration is not possible without a deliberate act involving the use of a tool or key, or should have a visible indication of their calibration or setting.

Overcurrent Protection Devices

Fuses

Type of fuses:

Semi-enclosed, often referred to as rewireable (BS 3036)

Cartridge (BS 1361) and (BS 1362)

High breaking capacity referred to as HBC (BS 88 Part 2)

High breaking capacity referred to as HBC (BS 88 Part 6)

Semi-Enclosed or Rewireable Fuses

Advantages of rewireable fuses

No mechanical moving parts

Cheap initial cost

Simple to observe whether element has melted

Low cost of replacing element

Disadvantages of rewireable fuses

Danger on insertion with fault on installation

Can be repaired with incorrect size fuse wire

Element cannot be replaced quickly

Deteriorate with age

Lack of discrimination

Can cause damage in conditions of severe short circuit

The diameter of copper wires to be used as fuse elements in these fuses is given in Table 53A of the Regulations.

Cartridge Fuses (BS 1361)

The body of the fuse can be either ceramic (low grade) or glass with metal end caps to which the fuse element is connected. The fuse is sometimes filled with silica sand.

Advantages

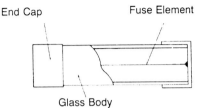

Small physical size

No mechanical moving parts

Accurate current rating

Not liable to deterioration

Disadvantages

More expensive to replace than rewireable fuse elements

Can be replaced with incorrect cartridge

Not suitable where extremely high fault current may develop

Can be shorted out by the use of silver foil

HBC Fuses (BS 88)

The barrel of the High Breaking Capacity fuse is made from high grade ceramic to withstand the mechanical forces of heavy current interruption.

Plated end caps afford good electrical contact.

An accurately machined element usually made of silver is shaped to give precise characteristic.

Some fuses are fitted with an indicator bead which shows when it has blown.

Advantages

Discriminates between overload currents of short duration, (e.g. motor starting) and high fault currents

Simple to observe when fuse has "blown"

Consistent in operation

Reliable

Disadvantage

Expensive

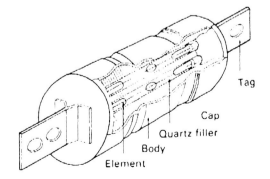

Circuit Breakers

Miniature Circuit Breakers (MCB's)

British Standard .

Miniature Circuit breakers should be manufactured to BS 3871

Types of MCB:

Thermal and magnetic

Magnetic hydraulic

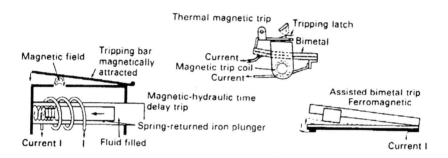

Assisted bimetal

Advantages

Tripping characteristic set during manufacture; cannot be altered

They will trip for a sustained overload but not for transient overloads

Faulty circuit is easily identified

Supply quickly restored

Tamper proof

Multiple units available

Disadvantages

Have mechanically moving parts

Expensive

Need for regular testing to ensure satisfactory operation

Classification of MCB's

MCB's and MCCB's are classified according to their instantaneous tripping current and the range of values are tabled for each type in BS 3871 Part 1. Type 1 is the most sensitive to tripping current of those listed. The type must be marked on the MCB when manufactured.

TABLE 2 BS 3871 Instantaneous Tripping Currents of MCB's

Type	Ampere			
1	>	$2.7\,I_n$	$\leq$	$4.0\,I_n$
2	>	$4.0\,I_n$	$\leq$	$7.0\,I_n$
3	>	$7.0\,I_n$	$\leq$	$10\,I_n$
4	>	$10\,I_n$	$\leq$	$50\,I_n$
B	>	$3\,I_n$	$\leq$	$5\,I_n$
C	>	$5\,I_n$	$\leq$	$10\,I_n$
D	>	$10\,I_n$	$\leq$	$20\,I_n$

Application of MCB's and MCCB's

To protect feeder cables

As an isolation point for a whole or section of an installation provided that contact clearances are adequate and contact position is reliably indicated.

Available for use on three phase and single phase supplies.

Differences between MCB's and MCCB's

The distinction between miniature circuit-breakers (MCB's) and moulded case circuit breakers (MCCB's) is on the basis of current rating and short circuit capacity.

BS 3871: Part 1 deals with units having current ratings up to 100A and short circuit capacities up to 9000A: such units have non-adjustable time/current characteristics with the mechanism sealed in a case.

Part 2 of the Standard includes units rated at 100A and above with breaking capacities in excess of 9000A. Not all of these larger units have a completely sealed enclosure for the circuit-breaker mechanism, although the trip mechanism must be sealed to prevent interference with the calibration except where adjustment is intended. These MCCB's are capable of handling greater prospective fault currents than MCB's.

Duty Ratings of Circuit-Breakers *(BS 3871 - Part 1)*

Category of duty	Prospective current of the test circuit (A)	Power factor of the test circuit
M.1	1000	0.85 to 0.9
M1. 5	1500	0.8 to 0.85
M.2	2000	0.75 to 0.8
M.3	3000	0.75 to 0.8
M.4	4000	0.75 to 0.8
M.6	6000	0.75 to 0.8
M.9	9000	0.55 to 0.6

Typical time current characteristics for a miniature circuit breaker (MCB)

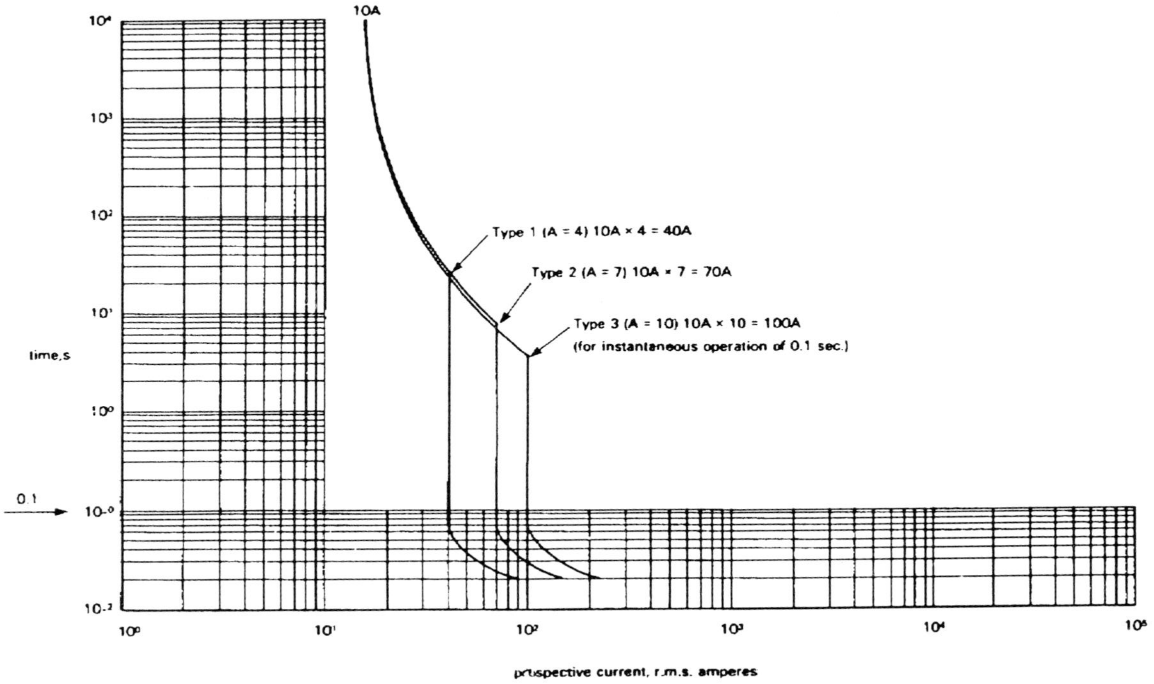

Type 1 (A = 4) 10A × 4 = 40A

Type 2 (A = 7) 10A × 7 = 70A

Type 3 (A = 10) 10A × 10 = 100A
(for instantaneous operation of 0.1 sec.)

10A

time, s

10^4

10^3

10^2

10^1

10^0

10^{-1}

10^{-2}

10^{-3}

0.1

prospective current, r.m.s. amperes

10^1

10^2

10^3

10^4

10^5

7B/4

Typical time current characteristics for a moulded case circuit breaker (MCCB)

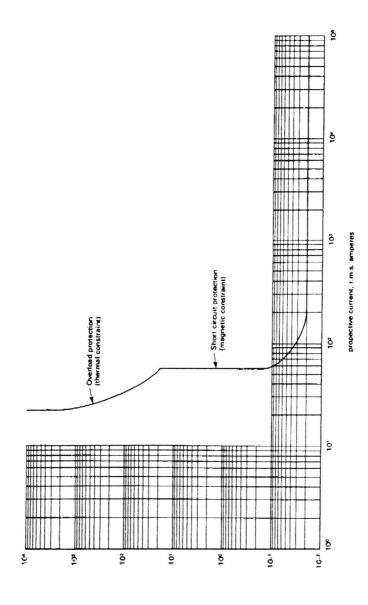

Overload protection (thermal constraint)

Short circuit protection (magnetic constraint)

prospective current, r.m.s. amperes

Overcurrent and Short Circuit Protective Devices

Operating Characteristics

Effects of a short circuit

- risk of causing a shock

- heat energy (proportional to I^2t)

- magnetic forces (proportional to I^2)

Operation of HBC Fuse Under Fault Current Conditions

An HBC fuse will normally operate under fault current conditions sooner than rewireable fuse or MCB, and an HBC fuse has the advantage of being totally enclosed and self-extinguishing.

The operating characteristics of an HBC fuse are its melting time and arcing time. The molten metal of the fuse element arcs and rapidly disperses into the silica sand, whereas a rewireable fuse or MCB takes longer to operate resulting in the production of greater heat energy and electro-magnetic stress.

Heat Energy

The illustration indicates the differences between the heat energy (I^2t) produced under short-circuit conditions when protection is afforded by an HBC fuse, as compared with a rewireable fuse or MCB.

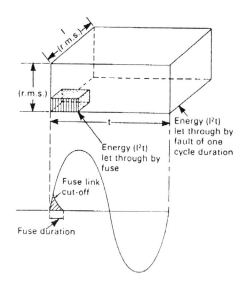

Magnetic Forces

The difference between the magnetic force (I^2) produced under short-circuit conditions when protection is afforded by a semi-enclosed fuse or MCB is compared with that of an HBC fuse; see illustration.

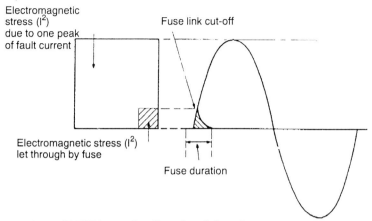

Operation of MCB's under Overload Conditions

Fuses give insufficient protection under conditions of small overload and operate too quickly under heavier overload conditions.

An MCB can detect a sustained overload of about 25%. An overload of 60-70% is needed to "blow" an HBC fuse and 100% overload is needed to "blow" a semi-enclosed (rewireable) fuse.

Discrimination

When designing a distribution system it is necessary to consider effective discrimination. Protective devices in an installation should be graded so that when a fault occurs the device nearest to the fault comes into operation. Other devices should remain intact.

Effective Discrimination

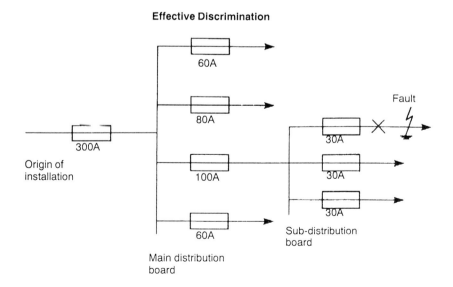

Operating characteristics of protective devices are given in manufacturer's literature to which reference is necessary to achieve effective discrimination in an installation.

Illustrated, is an I^2t characteristic for one manufacturer's HBC link. When total I^2t of the minor fuse in an installation does not exceed the pre-arcing I^2t of the major fuse, discrimination has been achieved.

I^2t characteristics - HBC fuse

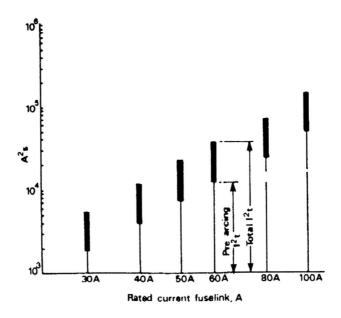

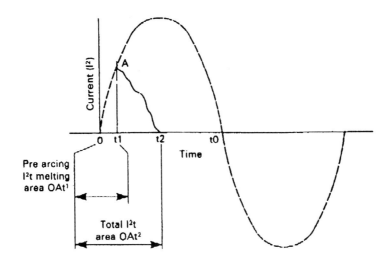

Selection of Live Conductors

Current Carrying Capacities of Cables
(Refer to Appendix 4, IEE Regulations)

The rating of a cable depends on its ability to dissipate the heat generated by the current it carries. This in turn depends in part on the type of installation. Table 4A (Appendix 4) of the IEE Regulations lists 20 standard methods of installation with examples identified by numbers. These classifications are use in the current carrying tables.

Example: From Table 4D2A of the IEE Regulations it can be seen that ordinary twin and multicore PVC-insulated cable installed in an enclosure such as trunking has a current capacity of only approximately 85% of what it would be when clipped to a surface or embedded directly in plaster.

Factors which effect the ability of a cable to lose heat (other than its physical characteristics) are:

- ambient (surrounding air) temperature

- cable grouping

Ambient Temperature

The rate of heat loss from a cable also depends on the difference in temperature between the cable and the surrounding air. A correction must be made to the current carrying capacity where the cable is to be installed in situations where high or low ambient temperatures may be expected to occur.

Tables 4C1 and 4C2 of the IEE Regulations give correction factors to be applied to the tabulated current-carrying capacities depending upon the actual ambient temperature of the location in which the cable is to be installed.

Cable Grouping

Cables installed in the same enclosure and all carrying current will get warm. Those near to the edge of the enclosure will be able to transmit heat outwards but will be restricted in losing heat inwards, while cables in the centre may find it difficult to lose heat at all.

The correction factors for this effect are given in Table 4B1 of the IEE Regulations. These relate to 'touching' and 'spaced' cables. 'Spaced' means a clearance between adjacent cables of at least one diameter.

Where the horizontal clearances between adjacent cables exceeds twice the overall cable diameter, no reduction factor need be applied.

It may well be that for a particular circuit the circumstances may change throughout its length, i.e. the ambient temperature or the number of cables bunched together may vary, or there may even be a change in the method of installation. Where this is the case the most onerous figure, i.e. that giving the LOWEST current carrying capacity must be used. Alternatively, the correction factors may be applied to the length of run specifically affected, but this will require the size of the cable to be increased for this length.

The correction factors for mineral insulated cables installed on perforated tray is covered in Table 4B2 of the IEE Regulations.

Notes

1. *The factors given in Table 4B1 and 4B2 of the IEE Regulations relate to groups of cables all of one size.*

2. *If due to known operating conditions a cable is expected to carry a current not more than 30% of its grouped rating it may be ignored when calculating the rating factor for the rest of the group.*

Determination of the Size of Cables

Having established the design current of a circuit and selected the type and size of protective device, it is necessary to determine the size of cable to be used.

Procedure

For BS 88 or 1361 fuse or circuit breakers to BS 3871 Part 2 or BS 4572 Part 2.

- Apply the correction factor for grouping of cables from Table 4B1 or 4B2 of the IEE Regulations.

- Divide the nominal current of the protective device (I_n) by any applicable ambient temperature correction factor given in Table 4C1 of the IEE Regulations for the type of insulation.

- Select the size of cable from the tables so that the current carrying capacity for the installation method is not less than the value of the protective device, adjusted as above.

If the protective device is a semi-closed fuse to BS 3036

- Divide In by 0.725 (not applicable to MI installations)

- Apply the correction factor for grouping from Table 4B1 or 4B2 of the IEE Regulations

- The nominal current of the protective device should be divided by the correction factors for ambient temperature in Table 4C2 of the IEE Regulations

The size of the cable selected should be such that its tabulated current carrying capacity is not less than the value of the protective device, adjusted as above.

In determining cable size, the correction factors are applied as divisors to the rated current of the overload protective device.

Example

A 240V lighting circuit consisting of 10 x 100 watt tungsten lamps is wired in PVC insulated single-core cable with copper conductors. It is protected by a 5A BS 3036 fuse. The cable is run through an ambient temperature of 35°C and is grouped with two other lighting circuits, e.g. two phase and two neutral conductors which are of the same size, and which are installed in the same conduit system on a wall. Determine the minimum cable size for compliance with the Regulation 523-01-01. The length of circuit run and voltage drop are to be neglected in this example.

Step 1

First check that the overcurrent protective device is adequate

$$\text{Current} = \frac{\text{Power (W)}}{\text{Voltage (V)}}$$

$$= \frac{10 \times 100}{240} = \frac{1000}{240}$$

$$= 4.2A \text{ (Satisfactory) This is the design current of the circuit } I_b$$

Apply the correction factors in the formula:-

$$I_t \geq \frac{I_n}{C_a \times C_g \times C_i \times 0.725}$$

Where
C_a = Correction factor or ambient temperature
C_g = Correction factor for grouping
C_i = Correction factor if cable is surrounded by thermal insulation
0.725 = If protective device is to BS 3036

Step 2

Determine correction factor (C_a) for the ambient temperature of 35°C from Table 4C2 of the IEE Regulations.

At 35°C correction factor with BS 3036 protective device = 0.97.

Step 3

Correction factor (C_g) from Table 4B1 of the IEE Regulations for these three circuits = 0.70.

Step 4

Correction factor for BS 3036 fuse = 0.725

Step 5

Determine minimum current carrying capacity of circuit live conductors

$$I_t \geq \frac{I_n}{C_a \times C_g \times 0.725} = \frac{5}{0.97 \times 0.70 \times 0.725} = 10.15A$$

Step 6

Select cable size from Appendix 4 Table 4D1A of the IEE Regulations (installation reference method 3 from column 4) 1.0 mm^2 = 13.5A current carrying capacity of conductor.

Thermal Insulation *(523-04)*

If a cable is to be run in a space in which thermal insulation is likely to be applied, the cable should be fixed wherever practical in a position where it will not be covered by the thermal insulation. Where this is not practicable, the cable CSA must be appropriately increased.

For cable in a thermally insulated wall or above a thermally insulated ceiling, where the cable is in contact with the thermally conductive surface on one side, the current carrying capacities are given in Appendix 4; reference method 4 of the IEE Regulations.

Where a single cable is likely to be totally surrounded by insulating material over a length of more than 0.5 m, in the absence of more precise information, the current carrying capacity is to be taken as 0.5 m times the current carrying capacity for that cable if clipped directly (reference method 1).

If a cable is surrounded by thermal insulation for less than 0.5 m, the current carrying capacity is reduced according to the cable size, length of run in insulation and the thermal properties of the insulation. Table 52A of the IEE Regulations gives derating factors for conductors up to 10 mm^2 in insulation having a thermal conductivity greater than 0.0625 W/Km.

Summary

When the overcurrent device is other than a semi-enclosed fuse to BS 3036 the cable selected has to be such that its tabulated current carrying capacity for the method of installation chosen is not less than:

$$I_t \geq \frac{I_n}{C_a \times C_g \times C_i} \quad \text{Amperes}$$

Where C_a = correction factor for ambient temperature
          C_g = correction factor for grouping
          C_i = correction factor if cable is surrounded with thermal insulation

Where the overcurrent device is a semi-enclosed BS 3036 fuse, a further correction factor has to be applied which is 0.725 (not applicable to MI cable installations).

$$I_t \geq \frac{I_n}{C_a \times C_g \times C_i \times 0.725} \quad \text{Amperes}$$

The steps to be followed when determining the size of cable to be used in a particular case are:

• Calculate the design current of the circuit

• Choose the type and rating of overcurrent device

• Divide this rating by any correction factors necessary

The result is the minimum current rating of the cable required, which is chosen from the Tables in Appendix 4 of the IEE Regulations.

✳ Voltage Drop *(525)*

A further consideration when selecting cables is that of volt drop. Regulation 525-01-01 requires that under normal service conditions the voltage at the terminals of any fixed current using equipment shall be greater than the lower limit required by the British Standard relevant to the equipment. Where the equipment is not the subject of a British Standard, the voltage at the terminals must be such as not to impair the safe functioning of the equipment.

The requirements of 525-01-01 are satisfied if on a supply given in accordance with the Electricity Supply Regulations 1988 the volt drop between the origin of the installation (usually the supply terminals) and the fixed current using equipment does not exceed 4% of the nominal voltage Uo.

Where an allowance has been made for diversity (Reg. 311-01-01) this may be taken into account when calculating voltage drop.

In certain circumstances, larger conductors than those permitted under Regs. 525-01 may be required to take the starting current of an electric motor. The effect of starting current on other equipment should also be taken into account.

Current carrying capacity tables (Appendix 4, IEE Regulations) also include values from which a cable volt drop can be calculated for each size of every cable type; the value given is expressed as a voltage drop per ampere per metre of cable (in millivolts).

To calculate the voltage drop, this figure must be multiplied by the length of the cable (in metres) and the design current on full load. The final product must be divided by 1000 to give the answer in volts.

Diversity must be taken into account when calculating volt drop. The application of rating factors means that in many cases the actual current is much less than the rated current, and the cable is cooler, and thus has a lower resistance than that calculated.

Calculation based on the tables will give a value for volt drop that is the highest possible figure. In marginal cases, an exact calculation, based on actual conductor resistance may be preferred.

Example

A 240V, 30A single-phase final circuit consists of a 22 m length of run of PVC insulated cable installed in conduit. The circuit has a design current of 26A. Determine the minimum size of cable which will comply with voltage drop requirements of Regulations 525-01.

$$\text{Maximum voltage drop allowed} = \frac{240 \times 4}{100} = 9.6V$$

Actual voltage drop

$$= \frac{mV/A/m \times design\ current \times length}{1000}$$

The simplest method to adopt is to determine the *maximum* mV/A/m which will comply with the *maximum voltage drop allowed,* in this case 9.6V.

$$\text{i.e. } 9.6V = \frac{mV/A/m \times 26 \times 22}{1000}$$

$$\frac{9.6 \times 1000}{26 \times 22}$$

$$\therefore \text{ max mV/A/m} = \frac{9,600}{26 \times 22}$$

$$= 16.78\ mV/A/m$$

By referring to Table 4D1B of the IEE Regulations any cable with 16.78 mV/A/m or less will therefore give an actual voltage drop of 9.6V or less when installed in accordance with the method to be used. For example, the cable is enclosed in conduit, using reference method 3.

Table 4D1B shows:

2.5 mm^2 = 18mV/A/m (hence v.d. greater than 9.6V)

i.e. $\dfrac{18 \times 26 \times 22}{1000}$ = 10.29V

4.0 mm^2 = 11 mV/A/m (hence v.d. **less** than 9.6V)

i.e $\dfrac{11 \times 26 \times 22}{1000}$ = 6.29V

The minimum size of cable which may be selected in this case is therefore 4.0 mm^2

Note: The above cable selection shows compliance only with Regulations 525-01

The following diagram is included to illustrate the voltage drop constraint on ring final circuits under different loading conditions.

For simplicity only half of the ring circuit has been shown.

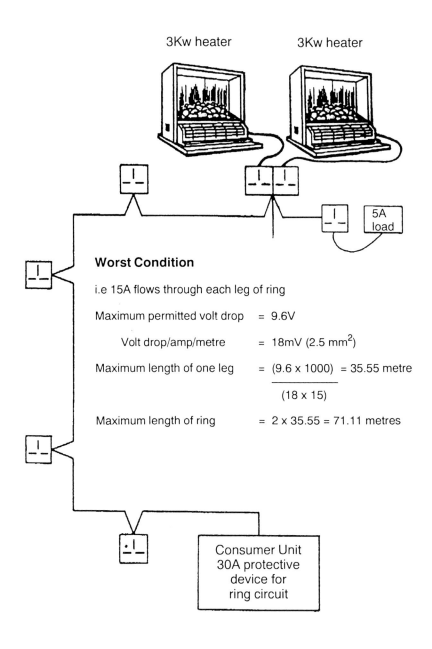

3Kw heater   3Kw heater

5A
load

Worst Condition

i.e 15A flows through each leg of ring

Maximum permitted volt drop = 9.6V

Volt drop/amp/metre = 18mV (2.5 mm^2)

Maximum length of one leg = $\dfrac{(9.6 \times 1000)}{(18 \times 15)}$ = 35.55 metre

Maximum length of ring = 2 x 35.55 = 71.11 metres

Consumer Unit
30A protective
device for
ring circuit

VOLTAGE DROP CONSTRAINTS
(30A Domestic Ring Circuits)

Sizing Conduit and Trunking Systems

Space Factors *(522 - 08)*

[✱] The number of cables which can be drawn in or laid in any enclosure of a wiring system must be such that no damage can occur to the cables or the enclosure during installation.

In order to comply with the above requirement a method employing a unit system is described in the IEE On-Site Guide, Appendix 5, where each cable is allocated a term. The sum of the terms for the cables which are to be run in the same enclosure is then compared with a factor given in the tables for different sizes of conduit or trunking, in order to determine the size of conduit or trunking necessary.

Wiring System

- Straight runs of conduit not exceeding 3 m (See IEE On Site Guide, Table 5B)

- Straight runs of conduit exceeding 3 m or run of any length incorporating bends or sets (See IEE On Site Guide, Table 5D)

- Trunking (See IEE On Site Guide, Table 5F)

Note: For conduit systems, a bend is classed as 90° and a double set is equivalent to one bend.

Cables

- Single core PVC insulated cables in straight runs of conduit not exceeding 3 m in length (See IEE On Site Guide, Table 5A)

- Single core PVC insulated cables in straight runs of conduit exceeding 3 m in length or runs of any length incorporating bends and sets (See IEE On Site Guide, Table 5C)

- Single core PVC insulated cables in trunking (See IEE ON Site Guide, Table 5E)

- Other sizes and types of cable in trunking, the space factor should not exceed 45%

Note: Only mechanical considerations have been taken into account in determining the factors in the tables. The values for cable capacities given in the Tables have been based on the application of an easy pull of cables into conduits.

As the number of cables increases in a conduit or trunking the current carrying capacity of the cables must be reduced in accordance with the application of the grouping factors (Table 4B1, IEE Regulations). In such cases it may be more economical to split the circuits and run them in more than one enclosure.

Example 1

A lighting circuit for a village hall requires the installation of a conduit system with a conduit run of 10 m with two right angle bends; the number of cables required is ten 1.5 mm^2 PVC insulated. What size conduit should be chosen for the installation?

Step 1

Select correct table for cable runs over 3 m with bends, from Table 5C, IEE On Site Guide

Step 2

Obtain term for 1.5 mm^2 cable - 22

Step 3

Apply term to number of cables in run = 22 x 10 = 220

Step 4

Select correct table for conduit systems runs in excess of 3 m with bends, from Table 5D, IEE On Site Guide

Step 5

Obtain from Table 5D a term for the length of run which is greater than 220. The table gives a term of 260 for a 10 metre run in conduit with two bends - 260.

Step 6

From Table 5D the conduit size is 25 mm - Answer: conduit is 25 mm

Note: For other cables and/or conduit consult manufacturer's information.

When other sizes and types of cable or trunking are used, the space factor should not exceed 45%. In this situation it is necessary to carry out the following procedure to determine the size of trunking required after the cable size has been decided.

- Consult the cable manufacturer's literature to obtain the overall dimensions of the cable including the insulation.

- Work out the cross sectional areas of the cables which are to be installed by using the formula:

$$\text{Cross sectional area} = \frac{\pi \times d^2}{4}$$

- Add together the individual cross sectional areas of the cables concerned and obtain the total cross sectional areas of the cables.

- Obtain the size of trunking by using the following formula which will allow a 45% space factor.

    let A = the c.s.a. of the trunking required

$$A \times \frac{45}{100} \text{ mm}^2 = \text{total c.s.a. of the cables}$$

$$\text{then } A = \frac{100}{45} \times \text{total c.s.a. of cables}$$

- Check manufacturer's trunking sizes and select one size. Convert to a c.s.a. and compare with calculated value.

Example 2

A steel trunking is to be installed as the wiring system for 8 single-phase circuits each having a design current of 35A.

40A BS 88 Part 2 fuses will be used as the overcurrent protective devices and PVC-insulated copper cables will be installed. Determine the size of trunking required.

Step 1

C_g is the grouping factor and from Table 4 B1 of the IEE Regulations the value for 8 circuits = 0.52

Minimum current carrying capacity of cables

$$= \frac{I_n}{C_g} = \frac{40}{0.52} = 77A$$

Using Table 4D1A of the IEE Regulations and installation method 8 the minimum size which can be used is 25 mm^2 cable.

Step 2

For 25 mm^2 cable the overall c.s.a. is 63.8 mm^2 (obtained from manufacturer's data).

Step 3

Total c.s.a. of cables = 16 x 63.8

$$= 1021 \text{ mm}^2$$

Step 4

Minimum trunking c.s.a. $= \dfrac{100 \times \text{total c.s.a. of cables}}{45}$

$$= \dfrac{100 \times 1021}{45}$$

$$= 2269 \text{ mm}^2$$

Step 5

From manufacturer's data the nearest trunking size greater than 2269 mm^2 is 50 mm x 50 mm (2500 mm^2)

10

Earthing Arrangements and Protective Conductors

Earthing Arrangements

General *(541)*

The earth can be considered to be a large conductor which is at zero potential. The purpose of earthing is to connect together all metal work (other than that which is intended to carry current) to earth so that dangerous potential differences cannot exist either between different metal parts, or between metal parts and earth.

Purpose of Earthing

By connecting to earth metalwork not intended to carry current, a path is provided for leakage current which can be detected and, if necessary, interrupted by the following devices

    — fuses
    — circuit breakers
    — residual current devices

 Where a building in which an electrical installation is being carried out is protected by a lightning protection system, account must be taken of the requirements of BS 6651 (Code of practice for protection of structures against lightning).

When bonding to a lightning conductor system, the bonding conductor must not have a cross-sectional area greater than that of the earthing conductor.

Connections to Earth *(542)*

The earthing arrangement of an installation must be such that:-

    • The value of impedance from the consumer's main earthing terminal to the earthed point of the supply for TN systems or to earth for TT and IT systems is in accordance with the protective and functional requirements of the installation and expected to remain continuously effective.

     • Earth fault and earth leakage currents which may occur under fault conditions can be carried without danger, particularly from thermal, thermomechanical and electromechanical stresses.

    • They are robust and protected from mechanical damage appropriate to the assessed conditions.

The installation should be so installed as to avoid risk of subsequent damage to any metal parts or structures through electrolysis.

Protection Against Indirect Contact

(Protection by earthed equipotential bonding and automatic disconnection of the supply) *(413-02)*.

General

Main equipotential bonding conductors should connect any extraneous conductive parts to the main earthing terminal of the installation. Extraneous conductive parts may include the following items.

- main water pipes
- main gas pipes
- other service pipes and ducting
- central heating and air conditioning systems
- exposed metallic parts of the building structure
- lightning protection systems

See local electricity company specifications for earth conductor sizes.

When an installation serves more than one building, the above requirements apply to each building.

Note: Bonding to metalwork of other services does require the permission of the authority responsible. Bonding to telelphone company earth wires should always be avoided unless specifically authorised by the telephone company.

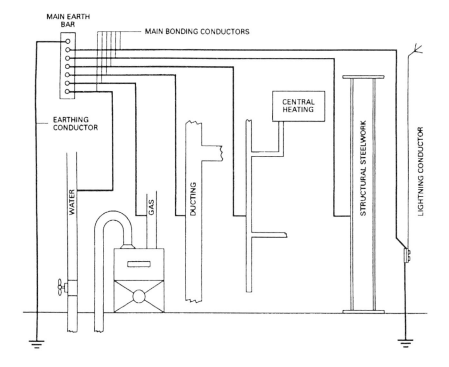

Equipotential Bonding

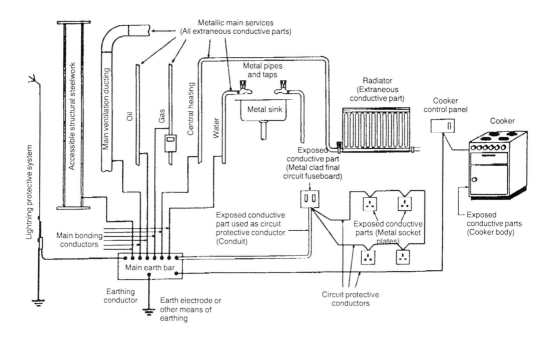

This type of bonding is intended to create a zone in which any voltages between exposed conductive parts and extraneous conductive parts are minimised. Simultaneously accessible exposed conductive parts must be connected to the same earthing system.

Where a number of installations have separate earthing arrangements (such as the separate wiring installations found in farm buildings), any protective conductor which is run between any two of the separate installations must be:

- capable or carrying the maximum fault current likely to flow for those installations

  or

- be earthed within one installation only and insulated from the earthing arrangements of any other installation, in which case the protective conductor must be earthed only at the installation containing the associated protective device.

It is essential that the characteristics of the devices used for shock protection, the installation earthing arrangements and the relevant circuit impedances be co-ordinated to avoid danger in the event of an earth fault *(413-02-04)*.

The above Regulation is considered to be satisfied if:

- for final circuits supplying socket outlets, the earth fault loop impedance at every socket outlet is such that disconnection occurs within 0.4 seconds (not applicable to reduced voltage circuits complying with Regulations 471-15, or when the alternative method covered by Regulation 413-02-12 has been used).

- for other circuits supplying only fixed equipment (excluding outdoors and in bath/shower rooms, agricultural, horticultural or construction sites) the earth fault loop impedance at every point of utilisation is such that disconnection occurs within 5 seconds.

- for final circuits supplying socket outlets the alternative method described in Regulations 413-02-12 may be used which involves limiting the impedance of the protective conductor of the circuit concerned to the main earth terminal and effectively extends the disconnection time in the event of an earth fault from 0.4 seconds to 5 seconds.

For circuits where protection is provided by an overcurrent device, the earth fault loop impedance (Z_S) for compliance with Regulations 413-02-09 must not exceed the value in Table 41B1 or Table 41B2 - 41D of the IEE Regulations as appropriate to the type of circuit and type of the protective device and its rated current.

Note: Tables 41B1, 41B2 and 41D are tabulated for circuits having a nominal voltage to earth (U_o) of 240V r.m.s. For other value of nominal voltages (U_o) the tabulated impedance values are to be multiplied by $\dfrac{U_o}{240}$

Regulation 413-02-16 and 17 states that when a residual current device is used in an installation which is part of a TN or TT system, the product of its rated residual operating current in amperes and the earth fault loop impedance in ohms shall not exceed 50(V).

The purpose of this Regulation is to ensure that in the event of an earth fault, sufficient residual operating current will be developed to cause disconnection within the permitted disconnection times of Section 413-02. It also ensures that when there is no earth fault the presence of earth leakage currents from equipment supplied by the circuit being protected will not cause excessive voltages to occur on exposed conductive parts and between these parts and extraneous conductive parts.

Installations which are part of a TN system *(413-02-06 to 17)*

All exposed conductive parts should be connected to the earthed point of the source of supply via protective conductors at the main earthing terminal as illustrated.

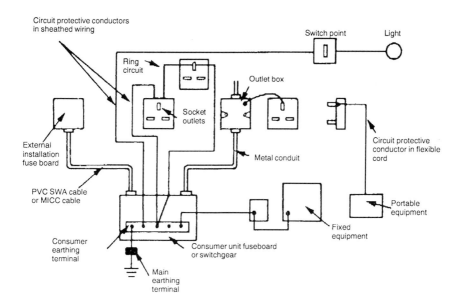

The following types of protective devices may be used in such an installation:

- Overcurrent

- Residual current - except where the neutral protective functions are combined in a conductor - (PEN conductor)

Installations which are part of a TT system *(413-02-18 to 20)*

The following types of protective device must be used in such installations:

- Residual current

- Overcurrent

Where protection is afforded by overcurrent or residual current devices all exposed conductive parts must be connected to an earth electrode or electrode system by protective conductors.

Installations which are part of an IT system
(413-02-21 to 26)

The following types of protective devices should be used in IT installations individually or together

- Overcurrent protective device
- Residual current

No live conductor of the installation may be directly connected to earth.

All exposed conductive parts should be earthed.

An insulation monitoring device must be provided to indicate the occurrence of a first fault from a live part to exposed conductive parts, or to earth. This device should automatically disconnect the supply or give an audible and/or visual signal.

It is essential that the first fault which occurs on such a system is eliminated as quickly as practicable.

After the occurrence of a first fault, conditions for disconnection of the supply should be as specified for TN and TT systems.

See maximum disconnection times, Table 41E, IEE Regulations.

Protection by Earth-Free Equipotential Bonding *(413-05)*

Earth free local equipotential bonding can only be used in an area *under effective supervision and where specified by a suitably qualified engineer.* Such protection requires that all simultaneously accessible exposed and extraneous conductive parts are connected together by bonding conductors BUT NOT TO EARTH.

This type of method of protection cannot be applied to entire installations and is difficult to co-ordinate with other protective measures used elsewhere in the installation.

Precautions must be taken to ensure that persons entering the equipotential location cannot be exposed to a dangerous potentiai difference, in particular, where a conductive floor insulated from earth is connected to the earth-free equipotential bonding conductor.

This type of protection is rarely used in this country and its application would typically be found in research and some electro-chemical process industries, where it is found necessary for personnel to be at the same potential as the process equipment.

Earth Electrodes *(542-02)*

The following items are recognised by the IEE Regulation as suitable earth electrodes.

- Earth rods or pipes
- Earth tapes or wires
- Earth plates
- Underground structural metalwork embedded in foundation
- Metallic reinforcement of concrete structures, except prestressed concrete
- Other suitable underground metalwork
- Lead sheaths or other metallic coverings of cables, where not precluded by Regulation 522-02-05

The metalwork and pipes of public gas or water services **should not** be used as a protective earth electrode.

Earth electrodes must be installed in such a way that their resistance does not increase above the required value through climatic conditions such as the soil drying or freezing, or from corrosion etc.

The lead sheaths and other metallic coverings of cables may be used as earth electrodes, provided they are in effective contact with earth, provided the consent of the owner of the cables is obtained and that he arranges to warn the owner of the electrical installation of any proposed changes to the cable that will effect its use as an earth electrode.

For details of types and methods of installing earth electrodes refer to Technical Data Sheet 10A.

Earthing Conductors *(542-03)*

Where earthing conductors are buried in the ground they must have a cross-sectional area not less than that stated in Table 54A of the IEE Regulations.

Aluminium and copper clad aluminium conductors should not be used for underground final connections to earth electrodes. All connections of earthing conductors to earth electrodes must be electrically and mechanically sound and fitted with a label stating "**Safety electrical connection - do not remove**".

Main Earthing Terminal bars *(542-04)*

A main earthing terminal or bar must be provided in an accessible position for every installation, for the connection of the circuit protective conductors, main bonding conductors, functional earthing conductors and any lightning protection system bonding conductors. Provision must also be made for disconnection of the main earthing terminal from the means of earthing to permit measurement of the impedances of the earthing arrangement.

The method of disconnecting the earthing terminal from the means of earthing must be such that it can only be effected with the use of tools.

Protective Conductors *(543-01)*

Cross-Sectional Areas

The minimum cross-sectional area of protective conductors can be obtained by using Table 54G (IEE Regulations) which establishes the minimum cross-sectional area of protective conductor in relation to the cross-sectional area of associated phase conductor. Alternatively the following formula may be used;

$$S = \frac{\sqrt{I^2 t}}{k} \quad mm^2$$

where:

S is the minimum cross sectional area in mm^2
I value of maximum fault in amperes
t operating time of the device in seconds
k factor for specific protective conductor
(Tables 54B-F of the Regulations)

Where a protective conductor is common to several circuits, the cross-sectional area should be calculated (Reg. 543-01-03) for the most unfavourable values of fault current and operating time in each of the circuits; or selected to correspond with the cross-sectional area of the largest of the various live conductors.

If the protective conductor does not form part of a cable, is not a conduit, ducting or trunking, and is not contained in an enclosure formed by the wiring system, the cross-sectional area should not be less than:

- 2.5 mm^2 if sheathed, or otherwise provided with mechanical protection

- 4 mm^2 where mechanical protection is not provided

When metal enclosures are used as protective conductors the cross-sectional area should be at least equal to the application of the formula stated in the Regulation 543-01-03.

The nominal cross-sectional area for metal conduits are as illustrated, based on the standard sizes given in British Standards.

Size	Nominal cross-sectional area mm^2	
	Light gauge	Heavy gauge
16	47	72
20	59	92
25	89	131
32	116	170

The nominal cross-sectional areas for metal trunking shown below are based on the standard sizes given in British Standards.

Size mm x mm	Nominal cross-sectional area mm^2
50 x 37.5	125
50 x 50	150
75 x 50	225
75 x 75	285
100 x 50	260
100 x 75	320
100 x 100	440
150 x 50	380
150 x 75	450
150 x 100	520
150 x 150	750

The values given in these tables should only be used if the joints made in either conduit or trunking systems do not reduce the nominal cross-sectional area.

Note: The value of resistance of steel conduit and trunking manufactured to BS 4568 and BS 4768 respectively must not exceed $5 \times 10^{-3} \Omega/m$ (0.005 Ω/m).

Certain manufacturers have also published resistance values for their conduit and trunking.

Types of Protective Conductors *(543-02)*

- PVC insulated single core cable manufactured to BS 6004 colour green/yellow

- PVC insulated and sheathed cable with an integral protective conductor manufactured to BS 6004.

- Copper strip

- Metal conduit

- Metal trunking systems ⎤

- Metal ducting ⎬ enclosures

- MICC - cable sheath ⎦

- Lead covered cable sheath

- SWA cable armourings

- An extraneous conductive part complying with Regulation 543-02-06

When the protective conductor is formed by a wiring system such as conduit, trunking, MICC, armoured cables or sheathed and insulated cables, a separate protective conductor must be installed from the earthing terminal of socket outlets to the earthing terminal of the associated box or enclosure.

The circuit protective conductor of final ring circuits which are not formed by the metal covering or enclosures of a cable should be installed in the form of a ring having both ends connected to the earth terminal at the origin of the circuit, e.g. distribution board or consumer's unit.

Exposed conductive parts of equipment should not be used to form part of protective conductors for other equipment, except if the exposed conductive part is a metal enclosed busbar trunking system or similar enclosures of factory built assembly, e.g. the case of a distribution fuse board, which may form part of a conduit system and hence be a part of the protective conductor circuit and which is constructed to satisfy the following requirements:

- The electrical continuity is achieved so as to afford protection against mechanical, chemical and electro-chemical deterioration

- The cross-sectional area is not less that that resulting from the application of the formula

$$S = \sqrt{\frac{I^2 t}{k}} \quad mm^2$$

- There is a provision for the connection of other protective conductors at tap-off points

- Flexible conduit cannot be used as protective conductors; an additional circuit protective conductor should be installed as illustrated below, or installed within the conduit but be accessible at the terminations *(543-02-01)*.

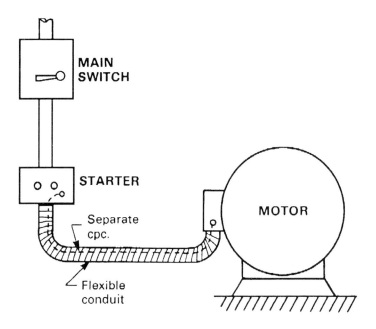

Preservation of Electrical Continuity of Protective Conductors *(543-03)*

Protective conductors should be installed so that they are protected against mechanical damage, chemical deterioration and electrodynamic effects.

Where protective conductors of cross-sectional areas up to and including 6 mm^2 are installed, which are not an integral part of a cable or a cable enclosure such as conduit or trunking, they must be protected throughout by insulation at least equivalent to that provided for a single-core non-sheathed cable of appropriate size complying with BS 6004 or BS 7211.

When the sheath of a cable containing an insulated protective conductor is removed for the purpose of making terminations or joints, protective conductors up to and including 6 mm^2 must be protected by a green/yellow insulating sleeve complying with BS 2848.

Connections of protective conductors should be accessible for inspection. This does not apply to (526-04):

— a compound filled/encapsulated joint

— joints made by welding, soldering, brazing or compression tools

— joints made in metal conduit ducting or trunking systems
(543-03-03)

All joints in protective conductors must be electrically and mechanically sound. Joints in metal conduit systems should be screwed or mechanically clamped: plain slip or pin grip sockets are not suitable as they will not ensure an effective low resistance joint throughout the life of the installation due to the small area of contact achieved with this type of fitting.

No switch should be inserted in a protective conductor.

Where an installation is connected to an electrical earth monitoring system, the operating coil of that system must be connected in the pilot conductor, not the protective earthing conductor (see illustration overleaf).

For further details of Earth Monitoring Systems refer to Technical Data Sheet 10B.

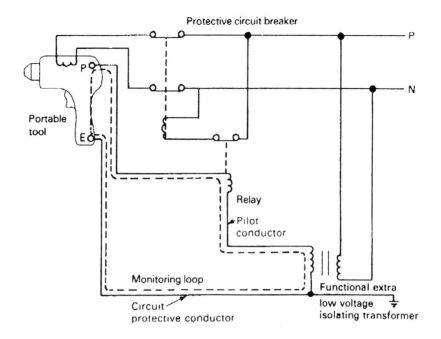

Protective circuit breaker

P

N

Portable
tool

Relay

Pilot
conductor

Monitoring loop

Functional extra

Circuit
protective conductor

low voltage
isolating transformer

Main Equipotential Bonding Conductor *(547-02)*

Main equipotential bonding to gas, water and other services should be made
as near as possible to the point of entry using bonding conductors with
cross-sectional areas of not less than half the cross-sectional area of the
earthing conductor; the minimum size being 6 mm^2. Except where PME
conditions apply, the cross-sectional area need not exceed 25 mm^2 if the
bonding conductor is of copper; or equivalent conductance where other
metals are used.

*Note: Electricity companies should always be consulted. Where PME instal-
lations are provided, check with supply authorities whether any special
requirements exist.*

Bonding of gas service pipes must be made on the consumer's side of the
meter between the meter outlet union and any branch pipework, but within
600 mm of the gas meter as illustrated on the next page.

If there is a water meter it should be shunted by a bonding conductor, to prevent damage to the working parts of the meter in the event of a fault current flowing through that section of pipework in which the meter is connected.

Note: Check with local electricity, gas and water companies for any special requirements regarding the bonding of services and main equipotential bonding conductor sizes. Otherwise use Table 54H of the IEE Regulations, which gives cross-sectional areas in relation to the copper equivalent cross-sectional area of the supply neutral.

Typical Example

Supply neutral conductor CSA	Main equipotential bonding conductor CSA
Up to 35 mm^2	10 mm^2
35 mm^2 to 50 mm^2	16 mm^2
50 mm^2 to 95 mm^2	25 mm^2

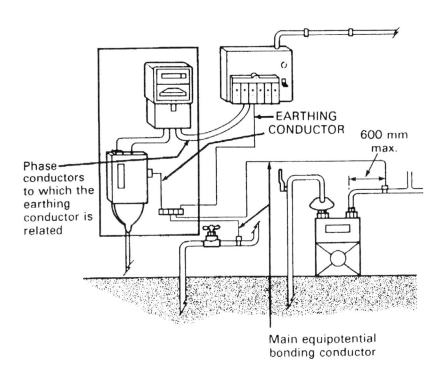

Main equipotential
bonding conductor

The purpose of installing bonding conductors is to ensure that any metalwork within an installation, such as gas and water services, are at the same potential as the metalwork of the electrical installation.

This is achieved by installing main bonding conductors from the main earthing terminal to the gas, water and other services at the point of entry to the premises as illustrated above.

Supplementary Bonding Conductors *(547-03)*

A supplementary bonding conductor used to connect exposed conductive parts must have a cross-sectional area not less than the smallest protective conductor connected to the exposed conductive parts, subject to a minimum of:-

- 2.5 mm^2 if sheathed or mechanically protected
- 4 mm^2 if mechanical protection is not provided

In situations where a supplementary bonding conductor connects two extraneous conductive parts, neither of which are connected to an exposed conductive part, the minimum cross-sectional area of the supplementary bonding conductor shall be:-

- 2.5 mm^2 if sheathed or mechanically protected
- 4 mm^2 if mechanical protection is not provided

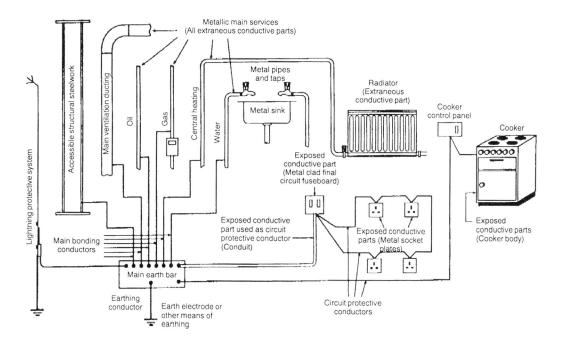

Supplementary bonding conductors may need to be installed in situations such as kitchens where a person may make simultaneous contact with an electrical appliance (such as an electric kettle or washing machine) and other metalwork, (such as the hot or cold water taps). In these situations supplementary bonding may be required as illustrated (but not if earth continuity tests prove that all metalwork of electrical, gas and water services is effectively bonded). The pipework of the gas and water services may be effectively connected together using permanent and reliable metal to metal joints of negligible impedance *(547-03-04)*.

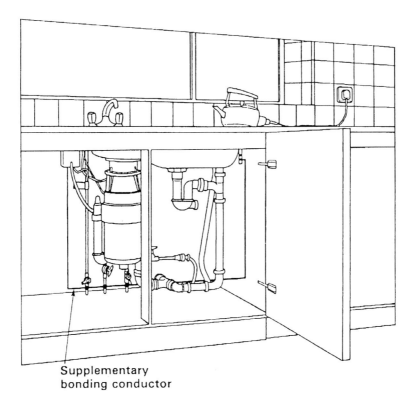

Supplementary
bonding conductor

A generally acceptable test which can be used to establish if an item is an extraneous conductive part is as follows.

Measure the insulation resistance using a 500 v d.c. insulation resistance test meter between the item and the main earthing terminal of the installation. If the value indicated on the test meter is greater than 0.25 megohm and a visual inspection confirms that this value is not likely to deteriorate, the item can, as far as is reasonably practical, be considered not to be an extraneous conductive part and does not require bonding.

In rooms containing a fixed bath or shower where permanent and reliable metal pipework has been installed, supplementary bonding conductors shall be installed as illustrated below, to reduce to a minimum the risk of electric shock in circumstances when the body resistance is likely to be low *(601-04-02)*. The most common method of making a bonding connection to pipework is by using BS 951 earth clips.

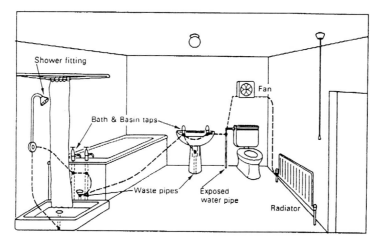

In other cases it is necessary to bond hot, cold and heating system pipework at one point only e.g. airing cupboard. This requirement does not apply to equipment supplied from a SELV circuit.

When electrical equipment is installed below the bath, eg. pumps for whirl-pool baths, the space must only be accessible by the use of a tool.

Combined Protective and Neutral (PEN) Conductors *(546-02)*

May only be used on systems where:

- The electricity company has been authorised in respect of the installation concerned, or

- The installation is fed from a privately owned transformer or converter with no metallic connection with the public supply system, or

- The supply is fed from a private generating plant

The types of cables which may be used as a PEN conductor are:

- conductors of cables not subject to flexing such as armoured PVC insulated cables and MICC cable, used on fixed installations having a cross-sectional area of not less than 10 mm^2 for copper conductors and 16 mm^2 for aluminium

- the outer conductor of concentric cables where the conductor has a cross-sectional area not less than 4 mm^2 in a cable manufactured to the appropriate British Standard

At every joint and termination point the continuity of the outer conductor of a concentric cable must be ensured by a conductor additional to any means of sealing and clamping the outer conductor. Illustrated below is a typical method of terminating an MICC cable for a two core used for a switch drop, e.g. switch feed and switch wire.

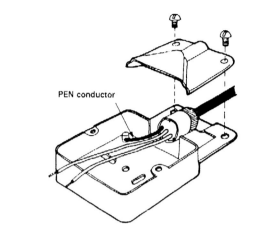

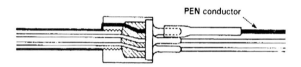

If, for any reason, the neutral and protective functions are provided by separate conductors run separately from any point in the installation, they must not be connected together anywhere beyond that point.

When MICC cable is used, the sheath must not have a cross-sectional area of less than 4 mm^2 and its resistance must never be greater than any of the internal conductors. This means that cables over 4mm^2 (which are classed as light duty) or 6 mm^2 (classed as heavy duty) cannot be used if they are single core cables, whose sheath cross-sectional area is always less than that of a single core (546-02-04).

The outer conductor of a concentric cable must not be common to more than one circuit. This requirement does not prevent the use of twin or multicore cable to serve a number of points contained within one final circuit.

No isolation or switching devices may be installed in the outer conductor of a concentric cable. The PEN conductor should be insulated so as to be suitable for the highest voltage present.

Note: Residual current devices must not be used to protect circuits where PEN conductors are to be installed.

For further information on PME supply systems refer to Technical Data Sheet 10C.

Types and Methods of Installing Earth Electrodes

Note: The information in this section has been obtained from manufacturers and the Code of Practice for Earthing, BS 7430: 1991.

Earth Plates

These are used where high fault currents are possible and normal soil conditions exist.

Typical manufacturers type and sizes are:-
(imperial measurements supplied by manufacturers)

Solid copper   2' x 2' or 3' x 3'
$\frac{1}{16}$" or $\frac{1}{8}$" thick

Lattice   2' x 2' or 3' x 3'
$\frac{1}{8}$" thick

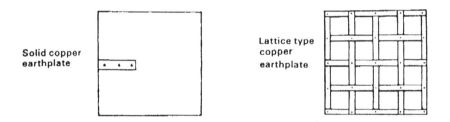

Solid copper earthplate

Lattice type copper earthplate

Method of Installing Earth Plates

Earth plates are buried vertically, which will reduce the effects of climatic conditions and extensive voltage gradients appearing on the ground surface under fault conditions. Care should be taken to ensure the electrode connection is protected against corrosion and mechanical damage.

Method of Terminating at Earth Plates

The following methods are used to connect earthing conductors to electrodes

- brazing
- tape connectors
- mechanical clamping devices
- aluminothermic welding or explosive welding techniques

Protection of Terminations

Earth electrode connections should be protected at the point of termination by the use of grease, bitumastic paint or bitumastic compounds.

Earth Electrodes for Area of Rocky-Soil Structure

In areas where there is rock at or near to the surface of the soil, copper tape, stranded conductors or wire mesh electrodes may be used.

These types of electrodes should be buried to a depth suitable to minimise the risk of their becoming damaged and to protect them from climatic conditions such as frost. (542-02-02). A suitable depth may be 457 mm.

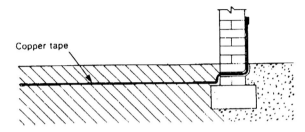

Methods of Installing Tapes or Stranded Conductors

Copper tapes and stranded conductors can be arranged in single lengths, or as parallel or radial groups. The usual size of copper tape is 25 mm wide by 3 mm thick.

Earth Rods

Earth rods are used where low earth fault currents exist and the soil resistivity high.

Types of Earth Rods

The following types of earth rods are available:

    — Copper earth rods
       Non-extensible 0.61 m and 1.22 m long with ribbed construction to provide maximum surface contact with soil.
       Extensible 1.22 m long, 15 mm diameter.

    — 'Biclad' earth rods
       Extensible 1.22 m long, 14.3 mm diameter made from copper sheathed mild steel bar.

    — 'Bimetal' earth rods
       Non-extensible 1.22 m long, 9 mm diameter complete with integral clamp for both strip and stranded conductors.
       Extensible 1.22 m. 1.83 m and 2.44 m rods, 15.9 mm diameter.
       'Bimel' rods consist of a corrosion resistant thick copper exterior, permanently molten-welded to a high tensile steel core.

Method of Installing Earth Rods

Earth rods are driven completely below ground using a hand-held hammer or power hammer.

They are set out at a distance of not LESS than their own length apart.

Earth Electrode Resistance Area

Every earth electrode has a definite electrical resistance to earth. Current flowing from the electrode to the general mass of earth has to traverse the concentric layers of soil immediately surrounding the electrode. Since the soil is a relatively poor conductor of electricity and as the cross-sectional areas of the layers of soil nearest to the electrode are small, the result is that of a graded resistance concentrated mainly in the area of soil surrounding the electrode. Moreover, the surface of the soil near the electrode will become "live" under fault conditions.

Surface Voltage Gradients

The illustration shows a typical surface voltage distribution near an earth electrode. The cow standing on the ground near the "live" electrode may receive a considerable voltage between its fore and hind feet resulting in a dangerous and possibly lethal shock since voltage of around 25V are dangerous to livestock.

Surface voltage gradients - danger to livestock

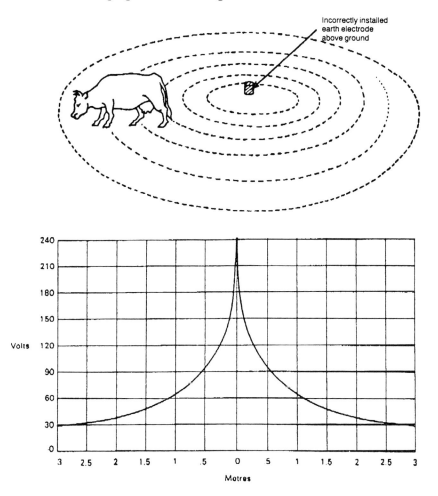

Termination to Earth Electrodes

The connection of earthing conductors to electrodes require adequate insulation where they enter the ground, to avoid possible dangerous voltage gradient at the surface. All electrode connections must be thoroughly protected against corrosion and mechanical failure.

It is important that the electrode is made accessible for inspection purposes, and a label should be fitted at or near the point of connection.

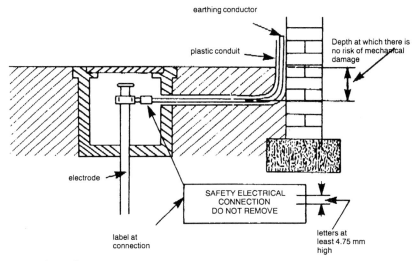

Inspection Covers

Constructed of concrete with a galvanised steel lifting handle, the underside is recessed to enclose and protect the electrode connection.

Standard pathway covers with cast iron frames with a concrete filling may also be used.

Glazed earthenware and fibre glass covers are also available.

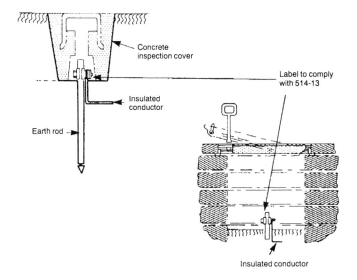

Earth Monitoring Equipment

Basic Earth Monitoring Units

The primary objective of a basic earth monitoring unit is to guard against the failure of the circuit protective conductor in a flexible or trailing cable. *BS 4444 1969 refers.*

Basic Circuit

This is as illustrated below and consists of the following:

- portable tool
- protective circuit-breaker
- transformer
- relay

The portable tool is supplied by a 4-core flexible cable consisting of:

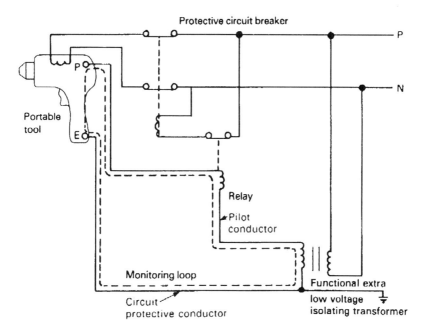

- The primary circuit protective which is bonded in the normal manner
- The pilot conductor
- The phase conductor
- The neutral conductor

The transformer is used to provide an extra low voltage supply of 12 volts and a current of not more than 3A around the loop circuit, consisting of:

-    the circuit protective conductor
-    a section of the metallic housing of the portable tool
-    the pilot conductor
-    the relay

Operation of the Relay and Circuit-Breaker

The circulating current holds in the relay and so energises the hold on coil of the circuit-breaker for as long as the earth monitoring loop remains intact.

Use of Earthed Screened Cables

Cable incorporating an earthed screening should be used in situations where vehicles or other equipment are likely to damage the cable. The phase conductor will become short-circuited to the earthed screening in the event of the cable being cut. This fault should then be cleared by an independent overcurrent device.

Protective Multiple Earthing (PME)

With this system the transformer neutral conductor is earthed and in addition the neutral is earthed at selected points in the distribution system. This has the effect of changing a phase to earth fault to a phase to neutral fault. When a PME installation is functioning correctly the degree of protection afforded is the same as that provided on a TN-S system.

Methods of Providing TN-C Systems (PME Supplies)

The following diagrams illustrate the methods used by area boards on PME distribution systems to provide supplies to consumers.

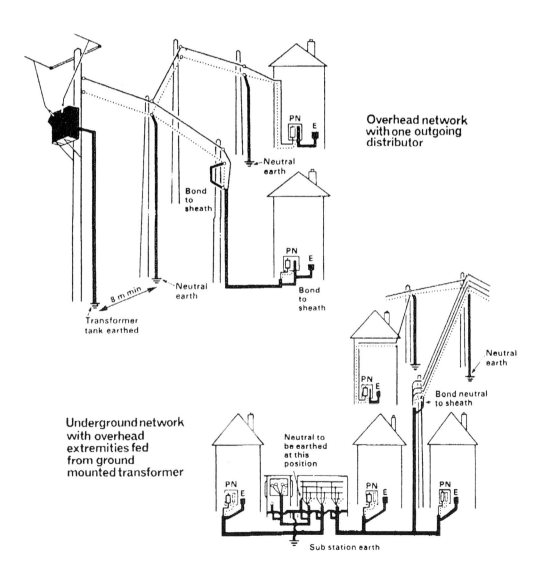

PNB Supply System

This is used where only one consumer is fed from a supply transformer on the network.

On construction sites and some agricultural installations it may not be possible to comply with the bonding requirements for PME approval. In these circumstances where a transformer supplies only one consumer, the electricity company will earth its neutral at one point only (at the service point) to provide an earthing terminal. This is known as Protective Neutral Bonding (PNB).

The method used to provide a PNB supply to consumer is illustrated below.

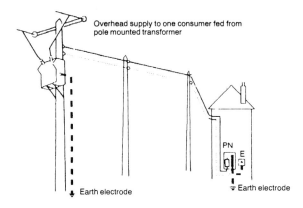

Typical TN-C-S Installation

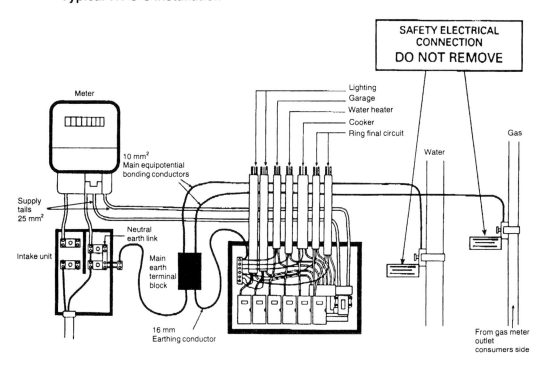

11

Cable Selection

Selecting Cables for Circuits and checking for compliance with Regulations

Certain tables and formulae used in this section are not part of the 16th Edition IEE Regulations. Reference has been made to the IEE On-Site Guide.

When installing a circuit, it is necessary to:

- Calculate the design current (I_b)

- Select the type and nominal rating of the protective device (I_n)

- Determine and apply correction factors to I_n

- Select cable from tables in Appendix 4, IEE Regulations (I_z)

- Calculate the voltage drop and check for compliance

- Check that circuit complies with shock protection

- Check that circuit complies with thermal constraints

EXAMPLE 1

A 20A radial socket outlet circuit is protected by a BS 88 Part II fuse. The circuit is ` ed using 2.5 mm^2 single core PVC cables installed in a 16 m length of PVC conduit. A separate protective conductor consisting of a 1.0 mm^2 PVC cable is used. Assuming that no rating factors are applicable and that the value of Z_E is given as 0.5 ohms, determine whether the circuit complies with the IEE Regulations. The nominal voltage (U_o) may be taken as 240V.

I_b Design current of circuit = 20A

I_n Protective device, BS 88 Part II fuse = 20A

Step 1

Apply correction factors for grouping and ambient temperature.

None apply

Step 2

Select suitable cable (Iz) from Appendix 4

from Table 4D1A, IEE Regulations, 2.5 mm^2 cable = 24A

Step 3

Calculate voltage drop

Maximum volt drop allowed = 4% of 240 = 9.6V

$$\text{Actual voltage drop} = \frac{\text{mV/A/m x design current Ib x length}}{1000}$$

$$\therefore \text{Actual v.d.} = \frac{18 \times 20 \times 16}{1000}$$

$$= \underline{5.76V}$$

Step 4

Check for compliance with shock protection

Total earth fault loop impedance = Zs

Maximum Zs from Table 41 B1, IEE Regulations = 1.85 ohms

$$\text{Actual Zs} = Z_E + (R_1 + R_2 \text{ ohms/metre x length})$$
$$Z_E = 0.5 \text{ ohms}$$

From Table 6A of the IEE On-Site Guide

(2.5 mm^2 phase conductor, 1 mm^2 protective conductor)

$R_1 + R_2$ = 25.51 milliohm/m

From Table 6B of the IEE On-Site Guide

Multiplier for PVC = 1.38

$R_1 + R_2$ = 25.51 x 1.38 x length of run

$$= \frac{25.51 \times 1.38 \times 16}{1000} = 0.56 \ \Omega$$

$$\text{Actual Zs} = 0.5 + 0.56$$

$$= \underline{1.06 \ \Omega}$$

Zs satisfactory i.e. 1.06Ω < 1.85Ω

Step 5

Check for compliance with thermal constraints

Calculate value of fault current $I_f = \dfrac{U_o}{Z_s}$

$$I_f = \frac{240}{1.06}$$

$$= 226.4A$$

Find value of t when $I_f = 226.4A$ from time/current characteristic for BS 88 20A fuse in Appendix, 3 Fig. 3A of the IEE Regulations

t will be less than 0.1 seconds, but for the purpose of the calculation use $t = 0.1$ secs.

Find value of k from Table 54C

$$k = 115$$

∴ minimum size of protective conductor $S = \dfrac{\sqrt{I^2 t}}{k}$ mm^2

$$S = \frac{\sqrt{226.4^2 \times 0.1}}{115}$$

$$S = 0.62 \text{ mm}^2$$

Nearest standard size of conductor = 1.0 mm^2. This will satisfy thermal constraints.

EXAMPLE 2

A 240V, 14kW domestic electric cooker installed in a house is to be supplied by a 15 m run of PVC insulated and sheathed cable clipped to a surface. The cooker control unit incorporates a 13A socket outlet.

The circuit is to be protected by a BS 3036 fuse and the value of external impedance is 0.85 ohms. Determine the minimum size of cable which may be used.

Step 1

Determine full load current

$$I = \frac{W}{V} = \frac{14,000}{240}$$

$$= 58.33A$$

Step 2

Determine design current I_b

Apply allowance for diversity from Table 1B, IEE On-Site Guide

Diversity for cooker = 10A + 30% of remaining current + 5A

design current $I_b = 10 + \dfrac{(30 \times 48.33)}{100} + 5$

$\qquad = 29.5A$

Step 3

Protective device I_n chosen

$\qquad = 30A\ BS\ 3036$

Step 4

Select cable size

$I_t \geq \dfrac{I_n}{0.725}$ (BS 3036 fuse protection factor = 0.725) $\therefore\ I_t \geq \dfrac{30}{0.725}$

$\qquad \geq 41.37$ Amperes

Select from Table 4D2A, IEE Regulations, Reference Method 1

$\qquad 6\ mm^2$ size = 46A

$\therefore$ Cable size to be used is $6\ mm^2$ phase with $2.5\ mm^2$ cpc

Step 5

Calculate voltage drop

Maximum v.d. = 4% of 240V

$\qquad = 9.6V$

Actual voltage drop $= \dfrac{mV/A/m \times I_b \times length}{1000}$

$\therefore$ actual voltage drop $= \dfrac{7.3 \times 29.5 \times 15}{1000}$

$\qquad = 3.23V$ *(satisfactory)*

Step 6

Check for shock protection constraints

from Table 41B1, IEE Regulations, Zs max = 1.14Ω

(Table 41B1 used because of socket outlet in cooker control unit)

actual Zs = Z_E + (R_1 + R_2) ohms

Z_E = 0.85Ω

(R_1 + R_2) = 10.49 milliohm/metre (from Table 6A, IEE On-Site Guide)

(Assuming 6 mm^2 phase conductor/2.5 mm^2 cpc)

Multiplier (for PVC) = 1.38 (Table 6B, IEE On-Site Guide)

$$\text{Actual Zs} = 0.85 + \frac{(10.49 \times 1.38 \times 15)}{1000}$$

$$= 1.067Ω$$

Check: actual Zs ≤ Zs max? *Yes - shock constraint satisfied*

Step 7

Check for thermal constraint

fault current If = $\dfrac{U_o}{Zs}$

$$= \frac{240V}{1.067}$$

$$= 224.9A$$

Value of time for 244.9A fault current for a BS 3036 fuse of 30A rating from the time current characteristic in Appendix 3, IEE Regulations, = 0.35 secs.

Value of k from Table 54C, IEE Regulations = 115

Minimum size of protective conductor S = $\sqrt{\dfrac{I^2 t}{k}}$

$$S = \sqrt{\frac{224.9^2 \times 0.35}{115}}$$

$$= 1.16 \ mm^2$$

Nearest standard size of protective conductor greater than 1.16 mm^2 is 2.5 mm^2 which is incorporated in the cable used.

EXAMPLE 3

An electric cooking appliance installed in a commercial premises is to be controlled by a unit fitted with a 13A socket outlet. The circuit is wired using a 12 m length of 6.0 mm^2 PVC sheathed cable clipped direct to a surface, and containing a 2.5 mm^2 cpc. The circuit is protected by a 45A, BS 1361 fuse and the value of Z_E for the installation is given as 0.45 ohms. Assuming that no rating factors are applicable for grouping or ambient temperature, determine whether the circuit complies with the Regulations. The normal voltage may be taken as 240V. The design current is 45A.

Design current I_b = 45A

Step 1

Protective device I_n = 45A BS 1361 fuse.

Check $I_n \geq I_b$? *Yes*

Step 2

Select size of cable from tables in Appendix 4, Table 4D2A of the IEE Regulations

Method 1

6.00 mm^2 = 46A (also referred to as Iz value)

Check $I_z \geq I_n$? *Yes*

Step 3

Calculate voltage drop

Maximum voltage drop allowed

$$= 4\% \text{ of } 240V = 240 \times \frac{4}{100}$$

$$= 9.6V$$

Maximum length of cable for compliance

$$= \frac{\text{Maximum volt drop}}{I_b \times \text{volt drop/A/metre}}$$

$$= \frac{9.6 \times 1000}{45 \times 7.3}$$

$$= 29.22 \text{ m}$$

(OK since this is greater than the 12m run required).

Step 4

Check for shock protection (413-02-04). Cooker classed as fixed equipment, but since socket outlet incorporated in control unit maximum disconnection time = 0.4 seconds.

from Table 41B1, IEE Regulations, Zs maximum = 0.6Ω

Method 1

Max $(R_1 + R_2) = Zs - Z_E$

$$= 0.6 - 0.45 = 0.15\Omega$$

from Tables 6A and 6B of the IEE On-Site Guide

$R_1 + R_2 = 10.49$ milliohm/metre x 1.38

$$= 0.0145\Omega$$

$\therefore$ max length $\dfrac{0.15}{0.0145}$

$$= 10.36 \text{ metres}$$

(does not comply since 12 m cable run required)

Method 2

Actual $Zs = Z_E + (R_1 + R_2$ milliohms/metre x length)

$Zs = 0.45 + (0.0145 \times 12)$

$$= 0.624\Omega$$

(too high; does not comply since max Zs = 0.6 Ω)

Question to be asked

Is socket outlet in control unit essential? If not, re-calculate value of Zs since circuit now only supplies fixed equipment and the maximum disconnection time for shock protection is 5 seconds.

From Table 41D (b), IEE Regulations, Zs max = 1Ω

from previous calculation for shock protection it can be seen that the circuit now complies for shock protection.

If socket outlet is essential try alternative method, (Table 41C, IEE Regulations)

Step 4

Alternative method Regulations 413-02-12 and 413-02-13

Method 1

Determine resistance/metre of 2.5 mm^2 cpc (R_2)

Table 6A gives values of resistance per metre and 6B gives multipliers to be applied (see IEE On Site Guide)

For 2.5 mm^2, resistance of cpc/m

$$= \frac{7.41 \times 1.38}{1000} = \qquad 0.0102 \ \Omega \ /m$$

Maximum length of cable $= \dfrac{Zs}{R \ of \ cpc \ / \ m}$

(from Table 41C) maximum cpc impedance $= 0.21\Omega$ for 45A BS 1361 fuse

Maximum length of cable $= \dfrac{0.21}{0.0102}$

$$= 20.59 \ m$$

Does comply, since the run required is only 12 metres

Method 2

$$Zs \ (actual) = \frac{(R_1 + R_2)/m \times CSA \ phase \ conductor \times length \ of \ run}{CSA \ phase \ conductor + CSA \ of \ cpc}$$

$$Zs = \frac{(0.0145 \times 6) \times 12}{8.5}$$

$$Zs = 0.1228\Omega \ OK \ since \ max \ value \ of \ Zs \ is \ 0.21\Omega$$

Note: Check for 5 seconds disconnection time from Table 41D, IEE Regulations

Zs max $= 1.0\Omega$

Zs actual $= 0.624\Omega$

Step 5

Check for compliance with thermal constraints

Fault current $I_f = \dfrac{Uo}{Zs}$

$$= \dfrac{240}{0.624}$$

$$= 384.6A$$

From time/current characteristic for BS 1361 45A fuse, (Appendix 3, Fig. 1, IEE Regulations)

t = 0.45 seconds for 384.6A

from Table 54C, IEE Regulations, k = 115

minimum size of protective conductor

$$S = \dfrac{\sqrt{I^2 \times t}}{k}$$

$$S = \dfrac{\sqrt{384.6 \times 0.45}}{115}$$

$$= 2.24 \text{ mm}^2$$

Nearest standard size is 2.5 mm^2 therefore cable with 2.5 mm^2 cpc is adequate.

12

Isolation and Switching (Chapter 46)

Requirements

The overall requirement is that every installation and all items of equipment should be provided with effective means to cut off all sources of voltage.

The terms for isolation and switching in the 16th Edition have specific meanings and these will be explained in this module.

Three main functions are dealt with:

- Isolation

- Switching off for mechanical maintenance

- Emergency switching

(There are also requirements for functional switching).

Where more than one of these functions is to be performed by a common device, the arrangement and characteristics of the device must satisfy all of the relevant regulations for the functions involved. The 16th Edition accepts that one device may be used to fulfil two or more of these functions which is often the case in practice.

An item of main switchgear required to disconnect all live conductors of a circuit must be designed to ensure that the neutral conductor cannot be disconnected before the phase conductors, and that the neutral conductor is reconnected before, or at the same time as the phase conductors. Regulation 476 specifies the alternative arrangements acceptable for a 4-wire, 3 phase a.c. supply. The protective conductor should not be disconnected. Where single pole switchgear is used it must not be connected in the neutral conductor of a TN, TT or any other system.

Note: In all cases when isolation for maintenance or alterations to circuits is being carried out, the isolator or switch must be fitted with a lock and label.

Isolation *(461)*

Isolation is defined as cutting off an electrical installation, a circuit or an item of equipment from every source of electrical energy in order to ensure the safety of those working on the installation by making dead those parts which are live in normal service.

Every installation must be provided with means of isolation, and attention needs to be given to the design, location and operation of any devices used.

An 'Isolator' is a mechanical device, operated manually, which is used to open or close a circuit when there is no load on the circuit.

Location and Operation of Isolation in Installations

The means of isolation must be located at a point as near as practicable to the origin of the installation. In the case of equipment, the isolator should be adjacent to the equipment it controls unless the requirements stated below for remote location are satisfied *(476)*.

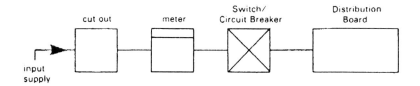

Switch should be located at origin of the installation

Where an isolator is used in conjunction with a circuit breaker as a means of isolating main switchgear for maintenance, the isolator must be interlocked with the circuit breaker so that it can be operated only when the circuit breaker is already open, or located where only an authorised and skilled person can operate it. *(476-02-01)*.

When isolating devices are to be placed at a location remote from the equipment to be isolated, one or both of the following arrangements must be made *(476-02)*.

(a) The primary means of isolation should be located so that only an authorised and competent person can operate it. It must be impossible for the isolator to be inadvertently returned to the 'ON' position; if a key or handle is used it should not be interchangeable with others used in the installation.

(b) A second means of isolation should be provided adjacent to the equipment.

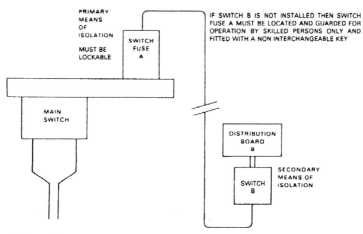

Remote location

Isolation of Motors *(476-02-03)*

Every motor circuit must be provided with an isolating device to disconnect the motor and all equipment, including any automatic circuit breaker.

Isolating and Switching Devices

Devices used for isolation must disconnect all live conductors (which by definition includes a neutral conductor) taking into account Regulation *460-01-02* and be installed so as to prevent unintentional reclosure (e.g. by mechanical shock or vibration).

The distances between the contacts of an isolator when open must not be less than that specified in BS 5419.

The position of the contacts of an isolator (i.e. open or closed) should be visible or clearly and reliably indicated *(461-01-05)*.

Before a person may work on an installation (or part of an installation) which has been isolated he must check that:

-   the isolator is OFF

-   the isolator cannot accidentally reclose

-   no other person can restore the supply

Semi-conductor devices must not be used as isolators. A "touch-switch" or a "photo-electric switch" are not suitable devices.

The choice of equipment selected as isolators is likely to be governed by the particular conditions of the installation. Suitable isolation devices are:

- Isolators

- Isolation switches (switch disconnectors) - not plate switches

- Links

- Fuse links

- Plugs and socket outlets

- Circuit breakers which comply with contact separation requirements

Some form of identification (clearly labelled) must indicate the circuit connected to the isolation device *(461-01-05)*.

Switching Off for Mechanical Maintenance *(462)*

A switch is defined as "A mechanical switching device capable of making, carrying and breaking current under normal circuit conditions which may include specified operating overload conditions and also of carrying for a specified time, currents under specified abnormal circuit conditions such as those of short-circuit".

A switch may also be capable of making, but not breaking, short-circuit currents.

Mechanical maintenance includes the replacement and cleaning of lamps, repair to heating elements which can be touched and work on electrical machinery. The switch or device is intended to prevent danger and possible physical injury. Special attention must be given to the location and operation of the device *(462)*.

The requirements are similar to those for isolation but generally the control switches will be local and must be capable of switching the full load current of the circuit or item of equipment *(537-03-04)*.

Devices Used for Switching Off for Mechanical Maintenance *(537-03)*

Devices must be selected and installed in such a manner as to prevent reclosure (e.g. by mechanical shock and vibration). Such a device must be capable of cutting off the full load current to the part of the installation it controls. It must have an external visible contact gap or a clearly indicated "OFF" or "OPEN" position. The device must only indicate "OFF" or "OPEN" when all poles of the supply have been isolated.

Devices which are suitable for switching off for mechanical maintenance include:

- Switches (excluding plate switches - Reg. 537-03-02)

- Circuit breakers

- Control switches operating contactors

- Plug and socket outlets

Whenever possible the switch must be inserted in the main supply circuit although it is permissible to insert it in the control circuit provided precautions are taken to provide a degree of safety equivalent to that of interruption of the main supply.

Devices for switching off for mechanical maintenance must be readily identifiable and convenient for their intended use.

Emergency Switching

Emergency switching involves the rapid disconnection of electrical energy to remove any hazard to persons, livestock, or property which may occur unexpectedly. Devices for emergency switching must be installed in a readily accessible position where the hazard might occur and it must not be possible for the supply to be restored from another location.

Plugs and socket outlets may **not** be used for emergency switching *(537-04-02)*.

Typically, emergency switching may be required in the event of a fire, accident, or an explosion.

There are additional requirements for discharge lighting installations operating at a voltage exceeding 1000 volts *(476-02-04)*. One or more of the following means must be provided:

An interlock on a self-contained luminaire arranged so that the supply is automatically disconnected before access to live parts can be made

- An effective means of isolation of the circuit installed near to luminaires

- Installation of a switch having a lock or removable handle, or a distribution board which can be locked

Note: For discharge lighting installations emergency switching must be in addition to the switch normally used for controlling the circuit.

Devices for Emergency Switching *(463)*

A device (switch or circuit breaker) used for emergency switching must be capable of cutting off the full load current to that part of the installation affected. Regulation 476-03 states that "a means of emergency switching shall be provided in every place where a machine driven by electric means may give rise to danger".

The device should be readily accessible and easily operated by the person in charge of the machine. Where more than one means of manually stopping the machine is provided, (e.g. remote stop-start devices), and danger might be caused by unexpected restarting, there must be a provision to prevent restarting.

Due account must be taken where stalled motor conditions may occur; in some cases it may be necessary to retain a supply to operate electric braking facilities, etc.

The means of emergency switching must consist of a single switching device which cuts off the incoming supply, or a combination of several items of equipment operated by a single action and resulting in the cutting off of the supply (e.g. sequential control). Such an arrangement with two contactors is illustrated below.

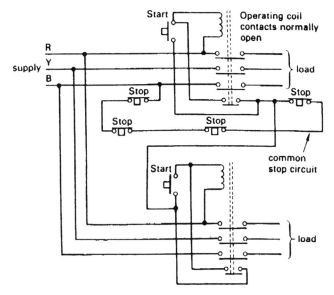

Emergency switching devices should, if possible, be manually operated, directly interrupting the main circuit. Where devices such as contactors or circuit breakers are operated by remote control, they should OPEN on de-energisation of the coil.

The operating means (i.e. handle or push button) should be clearly identifiable and coloured red, e.g. emergency stop buttons installed in laboratories and training workshops. They must be readily accessible and located where the hazard may occur. (Where appropriate at any remote operating position).

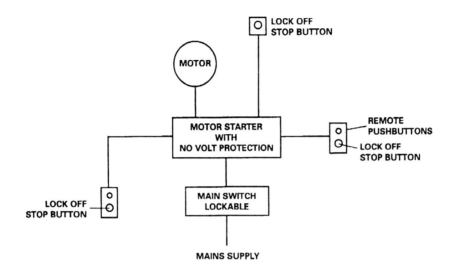

The switch must be of the latching type, capable of being held in the 'stop' or 'off' position. If the device is one which automatically resets itself, the operation and re-energisation of the device must be under the control of the same person. The release of any emergency operating device must not re-energise the equipment concerned unless suitable warning labels indicating that automatic restarting of the equipment may occur are clearly displayed. Plugs and socket outlets must not be used as emergency switching devices.

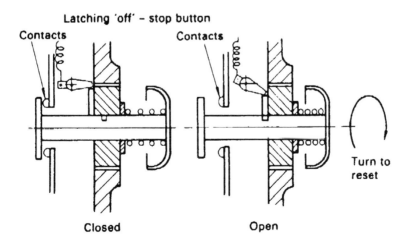

Emergency Switching by Means of a Fireman's Emergency Switch *(537-04-06)*

A fireman's emergency switch * is required for installations operating at more than low voltage (i.e. ≥ 1000V) in the case of:

- exterior discharge lighting

- interior discharge lighting when unattended

* Not applicable to a portable luminaire or sign of rating ≤ 100W and fed from a readily accessible socket outlet.

Applications:

A closed market or arcade is classed as exterior

A temporary installation to an exhibition in a permanent building is classed as interior

Location of a Fireman's Emergency Switch *(476-03-07)*

(a) For exterior installations the switch should be located outside the building and adjacent to the lamp or lamps (preferably one switch in any one building

(b) If not adjacent, labels should indicate the switch position

(c) For interior installations the switch should be in the main entrance to the building or in a position agreed with the fire authority; preferably one switch in any one building, independent of the switch for any external installation

(d) It should be conspicuous, reasonably accessible to firemen and usually not more than 2.75 m from the ground

(e) When more than one such switch is used, each switch must indicate the installation it controls, and the fire authority is to be notified.

Other requirements are that every fireman's switch should be coloured red, have fixed to it or adjacent to it a label with the words **"FIREMAN'S SWITCH"** clearly indicated. The label must be a minimum size of 150 mm x 100 mm with letters not less than 13 mm high, easily legible from a distance *(537-04-06)*.

The 'ON' and 'OFF' positions must be clearly indicated by lettering which can easily be read by someone standing on the ground. The switch must have its 'OFF' position at the top and be fitted with a lock or catch to prevent it being inadvertnetly returned to the 'ON' position.

Functional Switching *(537-05)*

The switching of electrically operated equipment in normal service is referred to as 'functional switching'. If this is not provided by a device which allows for switching off for mechanical maintenance or for emergency switching, then it is necessary to provide a switch to interrupt the 'on-load' supply for any circuit or appliance.

Devices of Functional Switching *(537-05-01)*

Plugs and socket outlets of rating not exceeding 16A may be used as a switching device (but not for emergency switching).

A plug and socket outlet exceeding 16A rating may be used as a switching device (a.c. only) provided it has a breaking capacity appropriate to the use intended.

Every switch for controlling discharge lighting must either:

- be designed for such a purpose in accordance with BS 3676

  or

- have a nominal current rating not less than twice the steady current which it is required to carry

  or

- where the switch controls both filament lighting and discharge lighting the switch must have a nominal current rating of not less than the sum of the current of the filament lamps, plus twice the steady current of the discharge lighting

- semi-conductor devices may be used for functional switching or control, provided they comply with the requirements of Section 512 of the Regulations

Other Requirements for Switching *(461 and 476)*

References, include the following for items not previously covered.

(a) A group of circuits may be switched by a common device (i.e. located between the mains intake and the consumer's main switchgear).

(b) Fixed or stationary appliances in a single room *(476-03-04)*. When two or more appliances are installed in the same room - as is often the case when a fitted kitchen incorporates built-in cooker, hob, microwave oven, fridge and dishwasher - then one single means may be used to control all the appliances e.g. a series of double pole switches in one control panel providing individual control of each appliance.

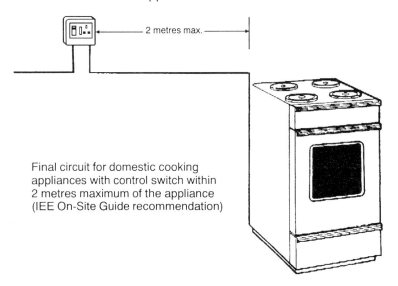

— 2 metres max. —

Final circuit for domestic cooking appliances with control switch within 2 metres maximum of the appliance (IEE On-Site Guide recommendation)

The means of control shall be capable of breaking the load and be in an accessible position which does not put the operator in danger.

Control of fixed and stationary appliances in domestic premises

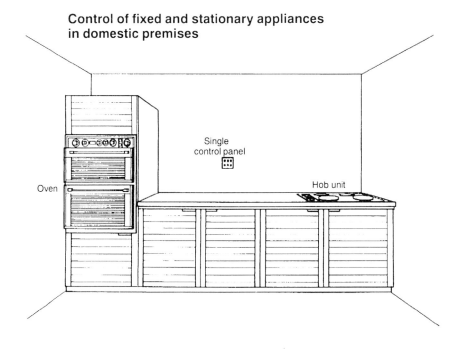

Oven

Single control panel

Hob unit

(c)   Every appliance of luminaire connected to the supply other than by means of a plug and socket must be controlled by a switch or switches separate from the appliance or luminaire and installed in a readily accessible position, subject to the provisions of Regulations *476-02-04*. When the appliance is fitted with heating elements which can be touched, the switch must be of the linked type which will break all circuit conductors, including the neutral (e.g. for a single phase supply a double pole switch *(476)*.

Note: For the purpose of the above regulations, the sheath of a silica glass sheathed element is regarded as part of the element which can be touched.

13

Protection Against Thermal Effects (Chapter 42)

The Regulations in this chapter are intended to deal specifically with the protection of persons, livestock, equipment and their environments from fire and burns.

Most Sections of the Regulations contribute in some way to protection against thermal effects, e.g. proper selection of type and size of cables, segregation, fire barriers, methods of installation, correct selection of over-current devices.

⁕ Protection Against Fire (422)

In the case of adjacent materials which may be a fire hazard or equipment installed where operating surface temperature may give rise to the risk of fire, one or more of the following methods must be adopted.

- Mounting on or enclosed in materials which will withstand the temperature generated, without the risk of fire or harmful effects

- Mounted to allow heat to dissipate safely

Special consideration must be given to enclosures for oil cooled transformers and switchgear when the quantity of liquid in a single location is in excess of 25 litres. The equipment should be housed with a drainage pit surround or in a chamber constructed of adequate fire resistant materials with sills to prevent the liquid spreading, and vented to the external atmosphere.

Fire Barriers

When any wiring systems pass through walls, floors or ceilings which are constructed of fire resistant materials, the hole should be filled and sealed with a material providing the same degree of fire resistance so as to prevent fire spreading from one part of a building to another (527-02-02).

Where wiring systems are installed in channels, ducts or ducting etc., which pass through a building, internal fire resistant barriers must be provided to prevent the spread of fire.

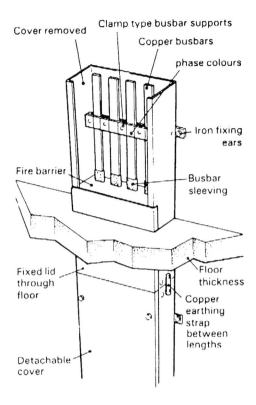

Fixed luminaires and lamps should be guarded to prevent ignition of any materials which are likely to be placed in close proximity. Shades or guards must be suitable to withstand the heat generated from the lamp or luminaire.

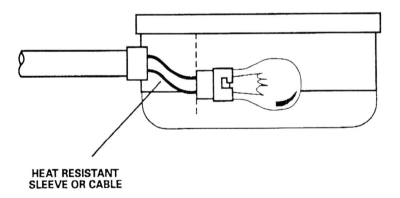

**HEAT RESISTANT
SLEEVE OR CABLE**

Protection Against Burns

⊠ Any part of an enclosure of fixed equipment liable to reach a temperature which may cause burns, must be located or guarded so as to prevent accidental contact. An exception is made for equipment manufactured to British Standards, specifically designed for use without a guarded enclosure.

Equipment within arms reach must comply with Table 42A of the IEE Regulations.

Typical Example

Metallic hand held equipment	55°C
Non-metallic hand held equipment	65°C
Metallic equipment which can be touched	70°C
Non-metallic equipment which can be touched	80°C

14

Selection and Erection of Equipment *(Part 5)*

Common Rules *(Chapter 51)*

General *(510)*

All items of equipment used in an electrical installation must be selected and erected so as to comply with the requirements of the IEE Regulations.

Compliance with Standards *(511)*

Every piece of equipment used in an electrical installation should comply with the requirements of the latest edition of the relevant British Standard or harmonised European Standard.

Note: A list of publications of the British Standards Institution to which reference is made in the IEE Regulations is given in Appendix 1 of the Regulations. For further details of BSI publications, reference should be made to the BSI Yearbook.

When a particular piece of equipment has been specified for a use which is not covered by a British Standard, the designer or person responsible for specifying the installation should verify that the equipment will provide the degree of safety required by the Regulations.

Operational Conditions and External Influences *(512)*

All equipment must be suitable for the

     — nominal voltage (rms value for a.c.)

     — design current (rms value for a.c.)

     — frequency

     — power characteristics

of the installation or part of the installation concerned.

Any equipment installed should be selected and erected so that it is compatible with all other equipment in the installation; and be of suitable design for the environment in which it is to function.

An example of a situation where external influences might affect the choice of equipment and conductors is a domestic dwelling, where the general temperature in the premises is unlikely to exceed 30°C, but may be considerably higher in airing cupboards or other localities. It is necessary to check that accessories, cables and other equipment affected are suitable for use at the higher temperature.

✳ Accessibility *(513)*

All equipment should be installed so that it can be easily operated, inspected and maintained and provide ease of access to any connections. This Regulation does not apply to joints in cables where such joints are permitted to be inaccessible. *(526)*

Identification *(514)*

All switchgear and control gear in an installation must be labelled to indicate its use. (This topic is covered in detail in Module 16, Identification Notices).

Mutual Detrimental Influence *(515)*

All electrical equipment should be selected and erected so as to avoid any harmful influences between the electrical installation and any non-electrical services; an example of a harmful influence to be avoided would be the heating effect created by leakage currents and resulting fire risk. When equipment carrying currents of different types (a.c. or d.c.) or at different voltages is grouped in a common assembly, all equipment using any one type of current or any one voltage must be effectively segregated from equipment of any other type, to avoid mutual detrimental influence, eg. the segregation of power circuits cables from telephone cables by use of multi-compartment trunking. *(515-01-02)*

Cables, Conductors and Wiring Materials *(Chapter 52)*

Types of Cable and Conductor

Non-flexible cables

- The following non-flexible cables are recognised as being suitable for low voltage electrical installations.

- Non-armoured PVC insulated cables to BS 6004, BS 6231 type B, or BS 6346.

- Armoured PVC insulated cables to BS 6346.

- Split-concentric copper conductor PVC insulated cables to BS 4553.

- Rubber insulated cables to BS 6007

- Impregnated paper insulated cables lead sheathed to BS 6480.

- Armoured cables with thermosetting insulation to BS 5467.

- Mineral insulated cables to BS 6207 part 1 or part 2.

Note: MI cables must not be used for discharge lighting circuits unless suitable precautions have been taken to avoid excessive voltage surges.

- Consac cables to BS 5593.

- Cables approved under Regulation 12 of the Electricity Supply Regulations. Paper insulated cables should comply to BS 6480 for non-draining cables where drainage of the impregnating compound is liable to occur *(521)*

Note: Aluminium cables with cross-sectional areas of 10 mm2 or less, cannot be used due to problems in achieving effective connection to wiring accessories.

The cables listed above can be used for overhead line wiring if the weight of the cable is supported by a catenary wire and the use is limited to low voltage installations. Where conductors are required for overhead lines the following types are recognised *(521).*

- Hard-drawn copper or cadmium conductors (BS 125)

- Hard-drawn aluminium and steel reinforced aluminium conductors (BS 215)

- Aluminium-alloy conductors (BS 3242)

- Conductors covered with PVC for overhead power lines (BS 6485 type 8).

Flexible cables

The following types of flexible cables and cords are recognised by the IEE Regulations for low voltage installations *(521)*

- Insulated flexible cords (BS 6500)

- Rubber-insulated flexible cables (BS 6007)

- PVC insulated flexible cables (non-armoured) (BS 6004)

- Braided travelling cables for lifts (BS 6977)

- Rubber-insulated flexible trailing cables for quarries and miscellaneous mines (BS 6116)

Types of Flexible Cables and Cords *(521)*

- Braided circular (PVC/copper braided)

- Unkinkable

- Circular sheath

- Flat twin sheathed

- Parallel twin*

  Note: Only for wiring of luminaires where permitted by BS 4533.

- Braided circular (Rubber/glass braided)

- Single core PVC insulated non-sheathed

For further details of non-flexible cables and flexible cables and cords refer to Appendix 3, IEE On-Site Guide.

Cables used on AC Circuits

Single core cables armoured with steel wire or tape should not be used on a.c. circuits, or single a.c. conductors installed in ferrous enclosures (such as conduit) unless the phase and neutral conductors are installed together in the same enclosure. The reason for this is that an alternating current induces an alternating magnetic field, the intensity of which is increased if the conductor is surrounded by a ferrous metal enclosure such as conduit or trunking; and this may result in eddy currents being circulated in the enclosure, causing a rise in temperature, giving rise to a fire risk. When the phase and neutral conductors are installed together in the same enclosure the magnetic field set up by the current in the phase conductor is effectively nullified by the equal and opposite magnetic field set up in the neutral conductor.

Conduits, Trunking, Ducting and Fittings *(521-05)*

These wiring systems must comply with the appropriate British Standard.

For conduit systems which are installed on site, each circuit should be completely erected before any cables are drawn-in. Where prefabricated conduit systems which include circuit cables are to be installed in a building, care must be taken to prevent the conduits and cables from being subjected to tension or mechanical stress or from being damaged during the construction of the building.

Busbar and Busbar Connections

All busbar and busbar connections used on electrical installations must conform to BS 5486 part 1.

Methods of Installing Cables and Conductors *(521-07)*

Cables can be installed by being either

- Enclosed in conduit trunking and ducting, or

- Open and clipped direct to non-metallic surfaces or bunched on cable tray, or

- Embedded in plaster, or

- Suspended from or incorporating a catenary wire.

Note: Table 4A of the Regulations

When cables are to be installed in ducts which have been cast in concrete the concrete surrounding the duct should be not less than 15 mm thick.

Operational Conditions *(523)*

Current Carrying Capacity

Cables selected should be capable of carrying the maximum sustained current for the circuits supplied (having considered the temperature and the environment in which the cable is installed)

Busbars, busbar connections and any bare conductors forming part of the equipment of any switchboard must comply with BS 5486 part 1 with regard to current carrying capacity and limitations of temperature. *(523-01-01)*

Except for a ring final circuit, cables connected in parallel to supply a load or piece of equipment, must be of the same construction, cross-sectional area, length and disposition, without branch circuits and arranged as to carry substantially equal currents.

When determining the current-carrying capacity of bare conductors, account must be taken of the arrangements made for their expansion and contraction, the joints and the physical limitations of the metal with which they are made *(523-03-01)*

Note: It is recommended that the maximum operating temperature of bare conductors should not exceed 90°C.

Busbars may operate at higher temperatures than most insulated conductors. When an insulated cable is to be connected to a busbar operating at a high temperature, the insulation of the cable must be checked to verify that it is suitable for the maximum operating temperature expected. Alternatively, the insulation and sheath should be removed for a distance of 150 mm from the point of connection and replaced as necessary by heat resisting insulation as illustrated.

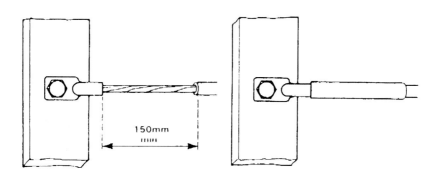

Insulation removed and replaced by High-temperature insulation

✳ Thermal Insulation (523-04)

Examples of cables run in insulation are given below

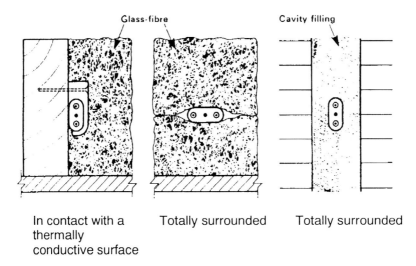

Glass-fibre Cavity filling

In contact with a
thermally
conductive surface

Totally surrounded

Totally surrounded

Where it is impossible to avoid running cables under the above conditions, a derating factor as illustrated below must be used. The current carrying capacity of cables likely to be totally surrounded by insulation for lengths greater than 0.5 m must be reduced by 0.5 (refer to IEE Regulations Table 52A). When a cable, installed in a thermally insulated wall or above a thermally insulated ceiling is in contact with a thermally conductive surface allowing heat to be dissipated on one side, the cable current carrying capacities given in Appendix 4, Method 4 (IEE Regulations) are to be used.

Derating Factors

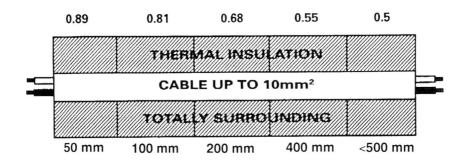

0.89	0.81	0.68	0.55	0.5

THERMAL INSULATION

CABLE UP TO 10mm²

TOTALLY SURROUNDING

50 mm	100 mm	200 mm	400 mm	<500 mm

Length in Insulation

Solid Bonding

Where single core metallic sheath or armoured cables are used, both ends of the cable run should be bonded together (solid bonding). However, this bonding might permit induced currents to circulate through the cable sheath and result in an increase in the temperature of the cable. The current rating tables in the regulations allow for this extra heat factor.

Where single core cables of 50 mm^2 or more are installed, the bonding at one end of the cable may be broken using an insulating plate. In these circumstances it may be possible to increase the current carrying capacity of the cable, but only after calculations have been made (by a qualified engineer) to confirm that the length of the single-end bonded cable is such that the induced e.m.f (which would drive a circulatory current if the cable were bonded at both ends) does not exceed 25V

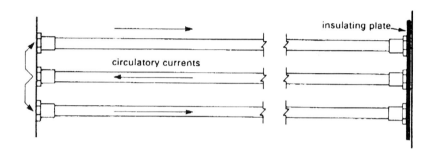

Single core cables, larger than 50 mm^2
Circulation of induced currents may be minimised if solid bonding is broken at one end by use of an insulator

Where single core cables pass through the metal plates of switches or busbars, the plate must be slotted to reduce the effects of eddy currents.

PLATE SLOTTED TO REDUCE EDDY CURRENTS

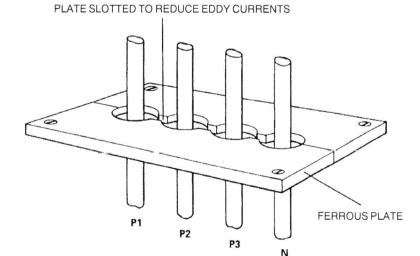

Voltage Drop *(525)*

The size of bare conductors or cables should be such that the voltage drop within the installation down not impair the safe function of connected equipment (in normal service). See relevant British Standard and design specification.

 The regulations will be satisfied if the volt drop between the origin of the installation and the fixed current-using equipment does not exceed 4% of the nominal voltage of supply.

Note: On installations supplying electric motors account should be taken of the effect of the motor starting currents on other equipment

Minimum Cross-Sectional Area of Neutral Conductors *(524-02)*

In cases where a three-phase load is made up of a number of single-phase loads, it may not be possible for the load overall to be balanced equally on the three-phase supply. The 'out-of-balance' current will return to the supply via the neutral conductor.

If a serious imbalance is unlikely to occur on the system then a multicore cable incorporating a reduced neutral conductor can be used. Where single-core cables are used in such circuits the neutral conductor should have a cross-sectional area which is capable of carrying the expected value of neutral current.

Where a circuit supplies discharge lighting the neutral conductor must have a cross-sectional area not less than that of the phase conductor.

Electromechanical *(521-03)*

All conductors and cables must be adequate for their purpose and installed so as to be able to withstand the electromechanical forces created by any current, including fault current, that will be carried by them.

Environmental Conditions *(522)*

✳ Ambient Temperature (AA)

The type of conductor, cable, flexible cord and joint used in the wiring of circuits, must be suitable for the highest operating temperature likely to occur in normal service. Account should be taken of the minimum temperature likely to occur so as to avoid the risk of mechanical damage to those cables susceptible to low temperatures, such as PVC insulated cables, which crack if installed in refrigeration plants where the temperature is consistently below freezing point.

Where cables or flexible cords are in contact with equipment or accessories which transfer heat, such as immersion heaters and luminaires, termination to this equipment should be made using heat resisting flexible cable or cord, or a suitable supplementary insulated sleeve or insulation should be applied to the conductors *(522-02)*

Enclosures should be selected so that they are suitable for the extremes of ambient temperature that will be encountered in normal service. A typical example might be PVC conduit which distorts in hot weather if expansion couplings have not been correctly fitted. Care must also be taken where non-metallic or composite outlet boxes are used for the suspension of luminaires to ensure they are suitable for the suspended load at the expected temperature *(522-02-02)*.

Where an installation consists of vertical runs of channel or trunking containing conductors or cables, internal barriers should be installed to prevent the warm air generated by the cables from rising to the top, which could lead to an excessively high local temperature creating a risk of breakdown of insulation. The barrier could also serve as a fire barrier, provided it was of a fire resistant material. Barriers should be installed between floors or every 5m, whichever is the less

Presence of Water or Moisture (AD AB) *(522-03)*

Wiring systems should not be exposed to rain, water or steam, but where this cannot be avoided, the wiring system should be selected to be corrosion resistant. PVC conduit and trunking, galvanised conduit etc., are normally used in these circumstances.

Care should be taken in damp and wet conditions to ensure that the composition of the wiring system and its fixings and accessories are chosen so that electrolytic action is prevented, e.g. an aluminium conductor should not be placed in contact with brass or copper. Copper clad aluminium conductors should not be used if these are likely to be exposed to damp or wet conditions.

Paper insulated and mineral insulated cables must be protected from moisture by being suitably sealed and terminated. Care must be taken to ensure that the insulation of MICC cable is dry before being sealed and terminated.

All conduit trunking and ducting systems should be installed so that the termination points of cables are protected from moisture and are corrosion resistant, and the access point of the system is placed so as to prevent the entry of water.

Where conduit systems are not designed to be sealed, the system must be provided with drainage outlets at any point where moisture is likely to collect *(522-03-02)*

Dust, Solid Foreign Bodies (AE) *(522-04)*

Enclosures installed in dusty environments such as woodworking shops, etc., must have a degree of protection to IP5X(BS 5490). (Protection to IP5X means that the equipment has been subjected to a test involving the use of talcum powder, which must not accumulate so that it could interfere with the correct operation of the equipment).

Corrosive and Polluting Substances (AF) *(522-05)*

Where a corrosive or polluting environment cannot be avoided the wiring system used should be of a type (or be protected) so as to withstand exposure to the corrosive or polluting substances. Non-metallic materials should not be placed in contact with oil/creosote or similar hydrocarbon substances likely to cause chemical deterioration.

Materials likely to cause corrosion of wiring systems:

- Materials containing magnesium chloride (used in construction of floors and dadoes)

- Plaster undercoats containing corrosive salt

- Lime, cement and plaster

- Oak and other acidic type woods.

- Dissimilar metals liable to set up electrolytic action.

Where conductors require termination involving soldering (as in the sweating on of cable lugs) the soldering flux used must not remain acidic or corrosive after the completion of the soldering process.

Fauna (AL) *(522-10)*

On farms and other situations where livestock is present, cables should be installed so as to be inaccessible to livestock and be either of a type resistant to attack from vermin or be suitably protected by being installed in conduit etc.

Solar Radiation (AN) *(522-11)*

Where cable and wiring systems are installed in positions which are exposed to direct sunlight, for example run along an outside wall of a building, the cable must be of a type which is resistant to damage by ultra-violet light. For cables the best protection is given by black sheathing compounds. If the cable is run out of doors it should also be weather resistant. In order to reduce the effect of solar radiation cables should be shielded from direct rays of the sun taking care not to restrict ventilation.

Installation Requirements

Mechanical Stress (AJ) *(522-08)*

Conductors and cables should be installed so that they are protected against any risk of mechanical damage.

Cables installed under floors and over ceilings must be run in such a way that they are undamaged through contact with the floor or ceiling or by the method of fixing. This involves careful routing and clipping of cables. Cables run in the space under floors and over ceilings should be installed at least 50 mm below the surface to prevent penetration by nails or screws used in fixing flooring and ceiling materials. Alternatively cables should be installed in an earthed steel conduit, securely supported, or provided with equivalent mechanical protection which will prevent penetration by nails or screws etc. *(522-06-05 and 522-06-07)*

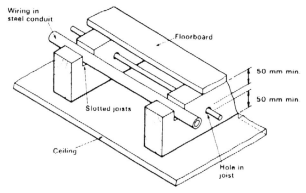

Support and protection for cables run under floorboards

Where cables pass through holes in metalwork, such as metal accessory boxes and luminaires, bushes or grummets must be fitted to prevent abrasion of the cables on any sharp edge.

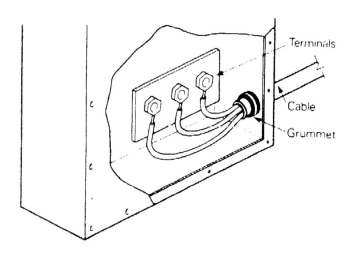

✱ Routing of PVC Cables

PVC flexible cables are always vulnerable to mechanical damage, particularly where they cannot be seen easily.

The practice of covering cables with plaster is widespread and cases do occur of nails and other objects penetrating cables and causing damage. This gives rise to the risk of electric shock. In practice this risk seems to be small, nevertheless it is desirable to reduce the risk as much as possible.

There are two methods of reducing this risk:

- Enclose the live conductors in earthed metal

- Place the cables where they are less likely to be damaged.

Regulation 522-06-07 requires the cable, when installed less than 50 mm^2 from the surface, to incorporate an earthed metal covering, or to be enclosed in conduit, trunking or the like. The enclosure must be substantial enough to fulfil the requirements of a protective conductor for the circuit in question, or by mechanical protection, sufficient to prevent damage to cables by nails or screws, etc.

Even heavy gauge steel conduit does not give complete protection against mechanical damage and the cable may need to be replaced. Capping of metal or plastic is used to protect cables laid under plaster. This will protect the cable during the plastering operation, but gives very limited protection against nails and other objects driven into the plaster.

New **Regulation (522-06-06)** requires cables, not protected as described in Reg 522-06-07 to be placed where they are less likely to be damaged. The permitted zones are as follows:

a) A strip 150 mm wide along the top of the wall and alongside an adjacent wall or partition.

b) A run either horizontally or vertically from the accessory to which it is connected.

In view of the practical problems in providing the earthed metal cover it is likely that cables will be installed mainly in the permitted zones.

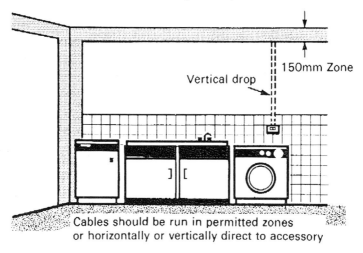

Cables should be run in permitted zones
or horizontally or vertically direct to accessory

Underground and Buried Cables

Underground cables installed in ducts, conduit or pipes should be insulated and incorporate a sheath or armour to give protection from damage, particularly when the cable is being drawn in *(522-06-01)*

Cables buried directly in the ground must be armoured or metal sheathed, or both, or be of the PVC insulated concentric type. Such cables must be marked by cable covers or marking tape and installed at a depth sufficient to avoid damage by disturbance of the ground. *(522-06-03)*

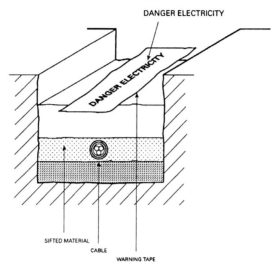

DANGER ELECTRICITY

SIFTED MATERIAL
CABLE
WARNING TAPE

Cable suitable for installation on outdoor walls are those incorporating a sheath and/or armour such as MICC or PVC SWA cable or those installed in a conduit system.

Overhead Cables

Where cables are installed overhead between buildings they should be installed out of reach of any source of mechanical damage, eg. vehicles, traffic, etc. Where access by vehicles is not possible the cables spanning between buildings may be installed in a heavy gauge steel conduit system without any join in the conduit *(521-01-01)*. See also 604-10 for construction site installations and Appendix 4 IEE On-Site Guide for cable supports.

Flexible Cords

Flexible cords should be selected in accordance with Table 4H3A, Appendix 4, IEE Regulations.

Flexible cords must not be used for fixed wiring unless contained in an enclosure providing protection against mechanical damage (521-01-03) or relevant provisions of the regulations are complied with.

Flexible cords should be used for making the electrical connection to portable or fixed equipment; care should be taken to keep their exposed length as short as possible without undue strain on the conductor or sheath, by making the connection to fixed wiring using suitable accessories such as plugs, sockets, joints, boxes etc. *(526)*

Note: Electric cookers with rated inputs of 3KW or greater are not considered to be portable.

Where luminaires are suspended from flexible cords the maximum weight supported must not exceed the value in Table 4H3A, IEE Regulations. For example; for 0.5 mm^2 flexible cord the maximum which can be supported would be 2 kg; for a 1 mm^2 flexible cord it would be 5 kg.

Identification *(514)*

All cores of cables and conductors should be identified at the points of termination and preferably throughout their length to indicate their function.

Methods of identification may include coloured insulation applied to conductors in manufacture or the application of coloured tape, sleeves or discs. The colours used must be those specified for the function in Table 51A and 51B

When armoured PVC insulated auxiliary cables, paper insulated cables or cables with thermosetting insulation are installed, an alternative identification system may be used using numbers, where 1,2 and 3 signify phase conductors and 0 the neutral conductor. The number 4 is used to identify any special purpose conductor.

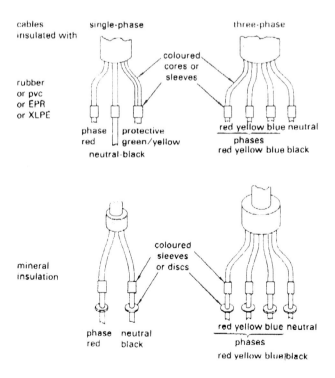

Flexible cables or cords should be identified throughout their length using the colours specified in Table 51B of the Regulations for the particular function of the core.

Note: The colour combination of green and yellow is to be used exclusively for the identification of protective conductors.

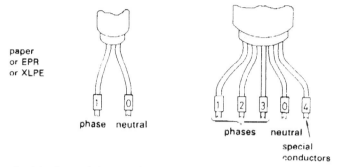

Where electrical conduits need to be easily distinguished from the pipelines of other services such as gas, oil, water, steam, etc. they should be painted orange.

Proximity to Electrical Services *(528-01)*

Low Voltage Circuits and Other Circuits

Where an installation comprises circuits for telecommunication, fire-alarm or emergency lighting systems, as well as circuits operating at low voltage and connected to a mains supply system, precautions must be taken to prevent electrical contact (and for fire alarm circuits and emergency lighting circuits, physical contact) between the cables of the various types of circuit.

Categories of circuits

Category 1 circuit – A circuit (other than a fire alarm or emergency lighting circuit) operating at low voltage and supplied directly from a mains supply system.

Category 2 circuit – With the exception of fire alarm and emergency lighting circuits, any circuit for telecommunications (e.g. radio, telephone, sound distribution, intruder alarm, bell and call and data transmission circuit) which is supplied from a safety source complying with Regulation 411-02-02

Category 3 circuit – A fire alarm circuit or an emergency lighting circuit.

⚹ *Note: Fire alarm circuits must be segregated from emergency lighting circuits 528-01-04*

The arrangements for enclosing cables of the three different circuit categories in an installation is illustrated below.

1 Trunking systems with separate partitions. Partition to be fire resistant if Category 3 cable installed.

2 Cables of the separate category circuit enclosed in their own independent conduit system

3 Cables of all three categories installed in the same enclosure provided that

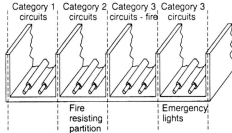

Circuit conductors of Category 2 are insulated to the same standard as Category 1 circuit conductors.

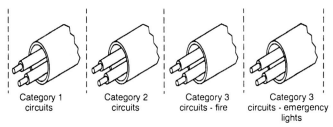

Category 3 circuit conductors are in mineral insulated cable.

Category 1 and 2 circuit conductors may be installed in the same enclosure, provided Category 2 circuit conductors are insulated to the same standards as Category 1 conductors and Category 3 conductors are installed separately.

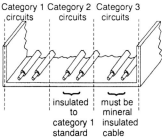

Where Category 1 and 2 circuit conductors are installed in conduit, ducting or trunking systems which terminate at boxes for the connection of accessories or controls, the two categories of circuit must be partitioned by means of rigidly fixed screens or barriers.

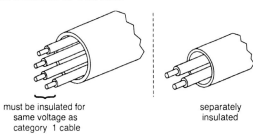

14/16

**Double depth intersection
(two compartment)**

In certain cases a single cable containing multiple cores, which are either insulated to the same standard or separated by an earthed braided screen may be used for the wiring of circuits in Categories 1 and 2.

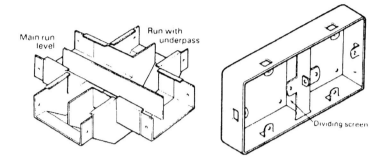

Proximity to Other Services *(528)*

Electrical Circuits and Exposed Metalwork

Metal sheaths and armour of all cables operating at low voltage, metal enclosures (conduit and trunking etc) and bare protective conductors associated with such cables should be installed so that no fortuitous contact occurs to other fixed metalwork; either by being effectively segregated or effectively bonded to it.

Electrical services should not be installed in the same conduit, ducting or trunking as pipes or tubes of non-electrical services e.g. air, gas, oil or water. This requirement does not apply where such services are under common supervision and it is confirmed that no mutual detrimental influences can occur *(528-02-04)*

No cables may be run in a lift (or hoist) shaft unless they form part of the lift installation *(528-02-06)*

Heat, Smoke or Fumes

⊠ A wiring system must not be installed in the vicinity of a service which produces heat, smoke, or fumes detrimental to the type of wiring used. (528-02-02). Eg. PVC cables close to balanced flue boilers.

Electrical Connections (526)

Every connection of a conductor must be:

— Mechanically sound

— Electrically sound

— Protected against vibration

— Protected against mechanical damage

— Terminated so there is no strain on the fixings of the connection

Methods of termination of non-flexible cables:

— Soldering

— Brazing

— Welding

— Mechanical clamps

All terminations of non-flexible cables must be:

— Made in a suitable accessory complying with British Standards appropriate to the size and type of conductor

— Suitably insulated for the voltage of the circuit

— Within an enclosure formed or completed with building material having ignitability characteristic P

Note: Where compression joints are used they should be made using the tools specified by the manufacturer and should be BS approved.

Methods of terminating wiring systems:

— Mineral insulated cable terminations having sleeves with a temperature rating similar to the seal

— Cable glands used for sheathed or armoured cable terminations shall securely retain the outer sheath or armour of the cable without damage

— Conduit ends should be reamed or filed to remove any burrs

Note: A material having ignitability characteristic P to BS 476 (Part 5) has been subjected to test where it has been placed in front of the flame of a gas jet for ten seconds. If the material does not flame and burning does not extend to the edge within this time, it is classified as 'not easily ignitable' and its performance indicated by the letter P.

Accessibility *(526-04-01)*

Every connection and joint must be accessible for inspection and testing and maintenance, with the following exceptions;

- A compound filled/encapsulated joint

- Connections between the cold tail and a heating element

- Joints made by welding, soldering, brazing or compression tool

Inspection-type conduit fittings where necessary in an installation, should be installed so as to be accessible for the purpose of removing or drawing cables.

⊞ Risk of the Spread of Fire *(527)*

Where a wiring system is required to pass through or penetrate material forming part of the construction of a building (eg cable trunking or busbar trunking systems), areas external to the wiring system, and where necessary, internal areas, must be sealed to maintain the required fire resistance of the material.

Wiring systems with non-flame propagating properties, having an internal cross section not exceeding 710 mm^2, need not be sealed internally.

The sealing system used must meet the following requirements;

- be compatible with the wiring system concerned

- permit thermal movement of the wiring system without detriment to the sealing

- be removable without damage when additions to the wiring system are necessary

- be capable of resisting external influences to the same standards as the wiring system

During the installation of wiring systems, temporary sealing arrangements must be made. In addition, any existing sealing which is disturbed or removed in the course of alterations to an installation, must be reinstated as soon as possible.

It is essential that sealing arrangements are visually inspected during installation to verify that they conform to the manufacturer's instructions. Details of those parts of a building sealed and the methods used must be recorded.

15

Special Installations and Other Equipment

Introduction

Special installations are now covered in separate sections of the Regulations; for example:

Bathrooms and showers Section 601
Swimming pools Section 602
Hot air saunal Section 603
Construction sites Section 604
Caravans and caravan parks Section 608
Highway power supplies and street furniture Section 611

Other equipment is dealt with in Chapter 55; for example:

Autotransformers
High voltage discharge lighting

Rotating Machines *(552)*

Equipment and cables feeding motors should be rated to carry the full load current of the motor. On starting, a heavy current will generate extra heat which will be dissipated in the cables for a short period of time. This will not cause overheating and should be ignored unless the motor is frequently re-started, in which case it may be necessary to install a cable of a larger cross-sectional area.

Machines such as wound rotor and commutator induction motors should be supplied by cables which are suitable to carry both starting and load currents.

Electric motors with ratings exceeding 0.37 kW must be supplied from a starter which includes overload protection. This requirement does not apply to a motor incorporated in an item of equipment complying with an appropriate British Standard *(552-01-02)*.

Every motor must be provided with a means to prevent automatic restarting after a stoppage due to a drop in voltage or failure of supply, where unexpected restarting of the motor could cause danger. For example, in a machine shop a failure of supply to a lathe could result in the operator making some adjustment or cleaning the machine which, if the machine restarts unexpectedly, could cause injury to the operator. In order to prevent this happening, 'no volt' protection should be provided. This is usually achieved by using a starter of the type illustrated overleaf.

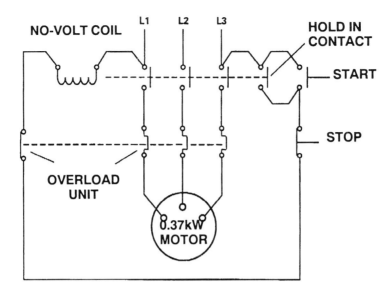

The coil is fed through hold-in contacts, which open when the button fails. The motor can only be restarted by operating the start button.

Note: The above requirement does not apply to motors where automatic starting is necessary, e.g. pumping systems, refrigeration and air conditioning units.

The metalwork of a rotating device may be used as a return conductor provided adjacent uninsulated metalwork is permanently and effectively earthed, and verified by testing.

Rooms Containing Fixed Baths and Showers
(601)

Socket outlets other than those in shaver units (BS 3535) must not be installed in a bathroom. Where a shower cubicle is located in a room other than a bathroom, (e.g. bedroom) all socket outlets should be installed more than 2.5 m from the shower cubicle (except shaver sockets to BS 3535), and be protected by an rcd (see section 8.1.6 IEE On-Site Guide).

SELV Installations

Socket outlets may be installed in a room containing a fixed bath or shower if wired on a SELV circuit not exceeding 12v RMS and complying with Regulations 411-01. The source for the SELV circuit should be installed out of reach of a person using the bath or shower, and the socket and switches used should have no accessible metal points and be protected against direct contact.

Supplementary equipotential bonding must be provided between exposed conductive parts of other equipment and extraneous conductive parts if simultaneously accessible. *For further details see Module 10, Earthing Arrangements and Protective Conductors.*

The protective device and earthing arrangements for circuits supplying fixed equipment should be such that in the event of an earth fault, disconnection occurs within 0.4 seconds *(413-02-09).*

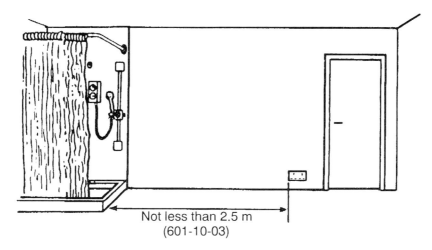

Not less than 2.5 m
(601-10-03)

Shower cubicle not in a bathroom

Lampholders installed within a distance of 2.5 m from a bath or shower cubicle must be constructed of, or shrouded in insulating material. Any bayonet type lampholders should be fitted with a protective shield; alternatively, a totally enclosed luminaire may be installed.

Switches or other electrical controls must be fixed so as to be inaccessible to a person using a bath or shower.

Not applicable to:

- switches supplied by a SELV system not exceeding 12V RMS a.c. or d.c.
- shaver units to BS 3535
- insulating cords or cord operated switches (BS 3676)
- controls incorporated in instantaneous water heaters to BS 3456
- mechanical actuators of electrical switches where an insulating linkage is incorporated in the mechanism

Any stationary appliance such as an infra-red heater or electric towel rail must be installed so as not to be within reach of a person using the bath or shower.

Wiring Systems

Metal conduit, trunking, exposed metal cable sheath surface wiring systems must not be used in locations containing a bath or shower *(601-07).*

Construction Site Installation *(604)*

The Regulations specifically cover this subject. See Section 604.

Aspects covered:

- New building
- Repairs, alterations
- Engineering construction
- Earthworks
- Similar works

This section does not apply to site offices, cloakrooms, meeting rooms, canteens, and toilets which are covered under the general requirements of the Regulations.

Supplies

25V	1ph	SELV	Portable hand lamps in damp and confined
50V	1 ph	CPE	Portable hand lamps in damp and confined
110V	1 ph	CPE	Portable tools and lighting up to 2kW
110V	3 ph	SPE	Small mobile plant up to 3.75kW
240V	1 ph		Fixed floodlighting
415V	3 ph		Fixed and moveable equipment above 3.75kw

C.P.E. = Centre Point Earthed
S.P.E. = Star Point Earthed

Regulation 604-12-02 states that plugs, cable couplers and socket outlets for use in construction site installations must comply with BS 4343, but this requirement need not be observed for installations in site offices and toilets.

Protection Against Indirect Contact

The voltage level to which any metalwork is allowed to rise under fault conditions is now reduced from 50 V to 25 V *(604-04-08)*. Therefore, the maximum disconnection times for installations connected to TN supply systems are reduced for a 240 V supply to 0.2 seconds. Refer to Table 604A. The revised maximum Zs values of loop impedance for the different types of protective devices corresponding to the disconnection times for each different supply system are given in Tables 604B1 and 604B2 of the IEE Regulations.

⊛ Caravans and Caravan Parks *(608)*

General

All the dangers which exist in every installation exist also in and around caravans, together with the problems which are created in moving a caravan, including the connection and disconnection from the source of supply.

Effective earthing is of vital importance, since in the event of a failure in the main protective conductor, a fault to the metal shell and chassis of the caravan could result in a shock to anyone touching earth.

Caravan Parks (Division Two)

The regulations require the following arrangements for caravan park installations.

- The earth terminal of every socket outlet supplying a mobile (touring) caravan must be earthed by one of the methods illustrated.

Mobile (touring caravans should be supplied from a socket outlet of the site installation. Each socket outlet must be supplied either singly, or in groups of not more than six socket outlets, through a residual current device with a residual operating current of not more than 30 mA *(608-13-05)*.

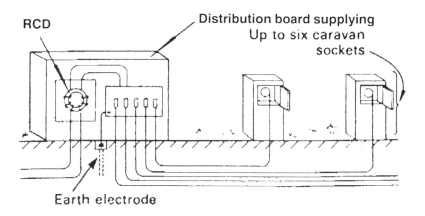

Every caravan must be supplied from a BS 4343 socket outlet (220/240V) which should be of the splashproof type IPX4 16A minimum current rating (608-13-02). Socket outlets must be located within 20 m of the caravan (608-13-01).

TN-S System

Building from which caravan supply is taken.

RCD

Underground cable.

Service cable.

Earthing conductor.

Socket outlet.

RCD at socket position

OR

See 608-07 Caravan couplers and inlet marking.

Duplicated protective conductor.

RCD

OR

RCD at socket position

Connection to the main earthing terminal by protective conductor unlikely to fail.

TT System

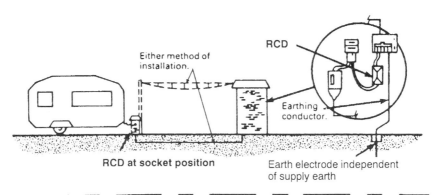

Either method of installation.

RCD

Earthing conductor.

RCD at socket position

Earth electrode independent of supply earth

TN-C-S System

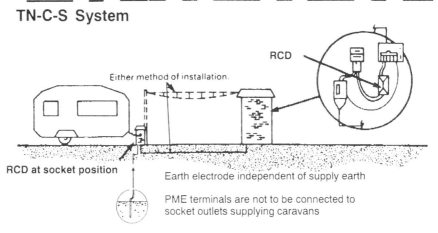

Either method of installation.

RCD

RCD at socket position

Earth electrode independent of supply earth

PME terminals are not to be connected to socket outlets supplying caravans

Caravans

The installation requirements for a touring caravan connected to a mains supply are illustrated below.

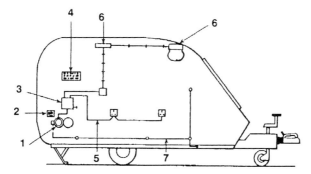

1. Provision of an inlet coupler to BS 4343 suitably recessed with lid.

2. Inlet coupler marked on the outside indicating the nominal voltage.

3. Protection, against indirect contact by automatic disconnection of the supply.

4. A notice installed adjacent to the main switch with instructions for electricity supply worded as Regulation *608-07-05.*

5. Protective conductors installed throughout each circuit within the caravan. Socket outlets must incorporate an earth contact. Class II equipment may be used.

6. Luminaires, preferably fixed directly to the structure or lining of the caravan. Where a pendant luminaire is installed, some means of securing it to prevent damage during transit must be provided. Enclosed filament lamps should be mounted so as to allow free circulation of air between the fitting and the body of the caravan.

7. Metal parts of a caravan must be bonded together and to the protective bonding conductor by a bonding conductor of not less than 4 mm^2 *(608-03-03).*

Note: If the caravan structure is fibreglass or other insulating material, the above requirement does not apply to any isolated parts, e.g. fixing brackets, which are not likely to become live in the event of a fault.

8. When sheathed cables are installed in inaccessible positions such as the ceiling or wall, the cable must be supported at intervals of 250 mm for horizontal runs and 400 mm for vertical runs.

9. The cross sectional area of every conductor must not be less than 1.5 mm^2.

Discharge Lighting *(554-02)*

High voltage electric signs and high voltage luminous discharge tube installations should be constructed, erected and installed to comply with BS 559.

Electrode Water Heater and Boilers
(554-04 and -05)

Electrode boilers and water heaters should only be connected to a.c.
systems and be controlled by a multipole linked circuit breaker, which when
operated disconnects the supply from all electrodes simultaneously. Overcurrent protective devices must be installed in each conductor feeding an
electrode.

The shell of any electrode boiler or heater should be bonded to the metallic
sheath or armour of the incoming supply cable with a conductor of cross-
sectional area not less than that of the largest phase conductor.

Electrode boilers or water heaters supplied directly from the supply at
voltages in excess of low voltage must be protected by a residual current
device which will disconnect the supply from the electrodes in the event of
a sustained earth leakage current of more than 10% of the rated current of
the boiler or heater. If this arrangement does not allow stable operation then
the setting of the residual current device for tripping may be increased to
15%, or a time delay may be included to prevent tripping due to short duration
transients.

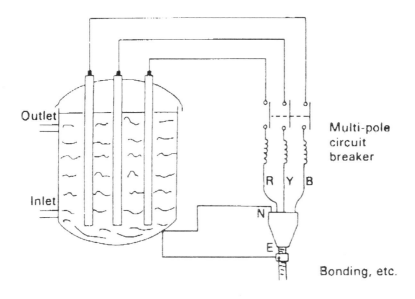

15/8

Electrode boilers and water heaters connected to a three-phase low voltage supply; the shell of the boiler or heater must be connected to the neutral conductor of the supply and the earthing conductor. The cross-sectional area of the neutral conductor should not be less than that of the largest phase conductor.

Single-phase electrode heater

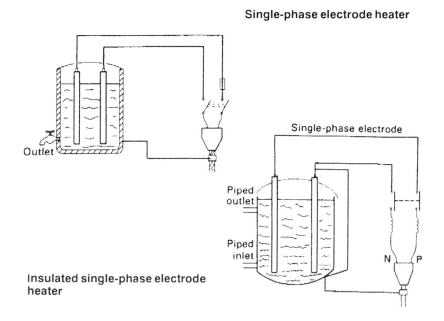

Single-phase electrode

Piped outlet

Piped inlet

Outlet

Insulated single-phase electrode heater

N   P

Single phase electrode boilers and heaters (which have one electrode connected to the neutral conductor earthed by the supply authority); the shell of the boiler or heater must be connected to the neutral of supply as well as the earthing conductor. Where the electrode boiler or heater is not piped to a water supply and where there is no contact with any earthed metal, and the electrodes and water in contact with the electrodes are shielded in insulating material so that it is impossible to touch either the electrodes or the water whilst the electrodes are live, a fuse may be substituted for the circuit breaker.

Every type of heater for water or liquids must incorporate (or be provided with) a thermostat *(554-04-01)*.

Water Heaters or Boilers with Uninsulated Heating Elements

All the metal parts of such boilers which do not carry current must be solidly and metallically connected to a metal water pipe through which the water supply to the boiler is provided. The water pipe must also be effectively bonded by an independent circuit protective conductor.

The supply to such heaters or boilers should be run through a double pole linked switch separate from, but within easy reach of the heater, the wiring from the heater or boiler being connected directly to the switch without the use of any plug or socket.

Soil and Floor Warming *(554-06)*

General

Soil and floor warming cables are used in a number of installations, such as:

- Installations under sports grounds to enable sports to be played even when frost and snow exists.

- Installations in roads and ramps of car parks to prevent icing and dangerous road conditions in winter.

- Installations under the soil in horticulture installations to enable plants and vegetables to establish early and healthy root systems.

- Installations in the floors of buildings used as storage type heating systems.

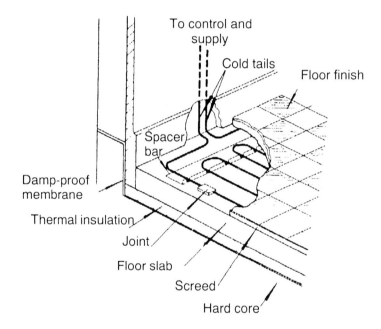

Where heating cables pass through or are in close proximity to materials which may present a fire hazard, the cables should be installed in fire resistant material with ignitability characteristic P (BS 476 Part 5) and be protected from mechanical damage *(554-06-01)*.

Heating cables intended for laying directly in soil, concrete, cement screeds or other materials used for the construction of roads and buildings should be:

-     Capable of withstanding any mechanical damage which is likely to occur during installation.

-     Constructed of materials that will be resistant to the effects of dampness and corrosion under normal service condition.

Heating cables installed in soil or roads or the structure of buildings must be:

-     Completely embedded in the material they are to heat.

-     Installed so as to prevent damage in the event of normal movements expected in the cables or substances in which they are embedded *(554-06-03)*.

The conductor temperature of floor-warming cables must not exceed that specified in Table 55C, IEE Regulations.

Agricultural Installations *(605)*

All electrical equipment in and around agricultural buildings should be of Class II construction or be constructed of or protected by suitable insulating material.

Where livestock have access to situations protected by SELV, the upper limits of the nominal voltage must be reduced as appropriate to the type of livestock.

When SELV is used, irrespective of nominal voltage, protection against direct contact must be provided by one of the following:

- Barriers or enclosure to IP2 X BS 5490.

- Insulation capable of withstanding 500 V a.c. rms. for 1 minute.

The very low body resistance of horses and cattle makes them susceptible to electrical shock at lower than 25 V rms. a.c. In agricultural installations where protection against indirect contact is provided by automatic disconnection, the limiting values of earth fault loop impedance are specified in the regulations. See Table 605 B1, IEE Regulations.

Protection Against Indirect Contact

The voltage level to which any metalwork is allowed to rise under fault conditions is now reduced from 50 V to 25 V. *(605-08-01)*. Therefore, the maximum disconnection times for installations connected to TN supply systems are reduced for a 240 V supply to 0.2 seconds. Refer to Table 605A. The revised maximum Zs values of loop impedance for the different types of protective devices corresponding to the supply system are given in Tables 605B1 and 605B2, IEE Regulations.

Electric Fence Controllers (605-14)

Mains-operated electric fence controllers must comply with BS 2632 or BS 6369 and be so installed that they are not exposed to the risk of mechanical damage or unauthorised interference. They must not be fixed to overhead power supply telephone poles except where the low voltage supply to the fence controller is carried by insulated overhead lines from a distribution board.

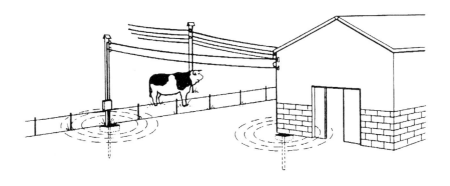

Earth electrodes connected to the earth terminal of an electric fence controller must be separate from the earthing system of any other circuit and should be situated outside the resistance areas of any electrode used for protective earthing. *For details of verification see Module 17, Testing and Inspection.*

Electric fence installations, including the controller and its conductors should be installed so they cannot come into contact with any power, radio or telecommunication systems (605-14-06).

Only one controller may be used to supply an electric fence installation.

Electric Surface Heating Systems (554-07)

Equipment for surface heating systems (ESH) must comply with BS 6351 (Part 1). The design of these systems should comply with Part 2 of the British Standard.

Installation and testing of the system should be carried out as provided for in BS 6351 (Part 3).

✳ Hot Air Saunas *(603)*

Irrespective of the nominal voltage, where SELV is used, protection against direct contact must be by one of the following:

- Insulation capable of withstanding 500 V a.c. rms. for one minute.

- Barriers or enclosures to IP 24 (BS 5490).

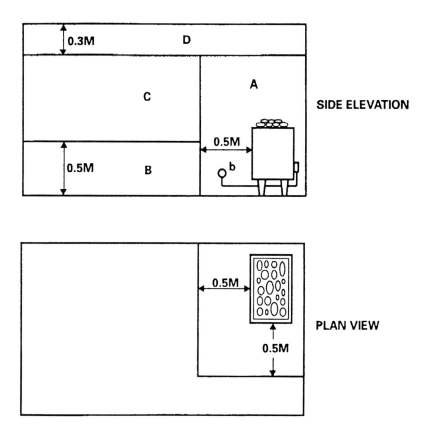

Zone A – Sauna equipment only

Zone B – No special requirement

Zone C – Equipment suitable for 125°C.

Zone D – Luminaires and sauna equipment suitable for 125°C.

✳ Swimming Pools *(602)*

This section applies to the basins of swimming pools, paddling pools, and surrounding zones. Special requirements may apply to medical pools.

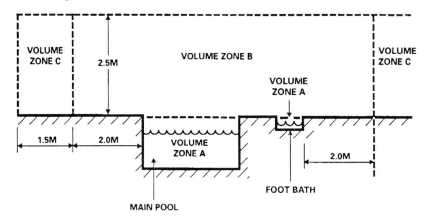

Protection Against Electric Shock

- Zone A & B; SELV max. voltage 12 V. Safety source outside zones A, B & C.

- Each floodlight is to be supplied from its own transformer.

- Automatic disconnection by means of rcd. for protection of socket outlets (412-06-02).

Equipment must have the following protection:

- Zone A – IP X 8

- Zone B – IP X 4

- Zone C – IP X 2 indoor pools

- Zone C – IP X 4 outdoor pools

- Swimming pools where water jets used for cleaning IP X 5.

Wiring Systems

Metal conduit, trunking and exposed metal cable sheath surface wiring systems must not be used for swimming pools *(602-06-01)*.

Zones A and B should only contain wiring to supply appliances installed in those zones *(602-06-02)*. No junction boxes shall be used *(602-06-03)*.

In Zone C, socket outlet, switches and accessories are permitted if protected:

    – individually by electrical separation
    – by SELV
    – by rcd with characteristics specified in Reg. 412-06-02
    – a shaver socket to BS 3535

Highway Power Supplies and Street Furniture *(611)*

Protection Against Shock *(611-02)*

In accordance with the requirement to protect against direct contact, an intermediate barrier must be provided to prevent contact with live parts, and afford protection to at least IP2X.

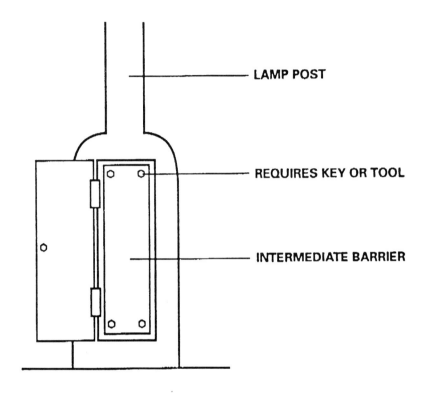

LAMP POST

REQUIRES KEY OR TOOL

INTERMEDIATE BARRIER

For circuits feeding fixed equipment used in highway power supplies and complying with Regulation 413-02-04, the maximum disconnection time is 5 seconds *(611-02-04)*.

When protection against indirect contact is provided by earthed equipotential bonding and automatic disconnection of the supply, any metal structures not forming part of or connected to street furniture or street located equipment must not be connected to the main earthing terminal as an extraneous conductive part; eg. metal pedestrian barrier to lighting column *(611-02-05)*.

When a programmed inspection and testing procedure is being operated the labelling requirements for inspection and testing in Regulation 514-12 need not apply.

✱ Restrictive Conductive Locations *(606)*

This section deals with installations either within, or supplying equipment or appliances within restrictive locations.

- Boiler de-scaling

- Metal factory gantry

- Steel structures

Where SELV or FELV is used for protection against electric shock, the voltage must not exceed 25 V. Barriers or enclosures must be to at least IP2X and insulation must withstand a test voltage of 500 V rms. for 1 minute.

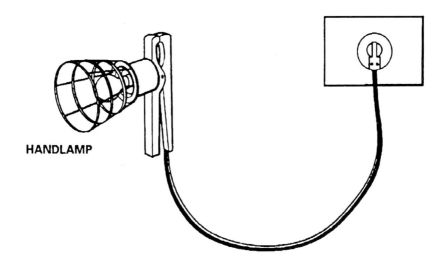

HANDLAMP

PROTECTED BY S.E.L.V. MAX VOLTAGE 25V rms AC

Identification Notices *(514)*

Switchgear Control Gear and Protective Devices

Switchgear and control gear in an installation should be labelled to indicate its use. Where the operation of switchgear or control gear cannot be seen by the operator, an indicator light or other signal should be installed. British Standard 4099 deals with the colours which should be used for indicator lights, push buttons, annunciators and digital readouts.

All protective devices in an installation should be arranged and identified so that their respective circuits may be easily recognised.

Diagrams *(514-09)*

Diagrams and charts must be provided for every electrical installation indicating:

(a) the type of circuits

(b) the number of points installed

(c) the number and size of conductors

(d) the type of wiring system

(e) the location and types of protective devices and isolation and switching devices

(f) details of the characteristics of the protective devices for automatic disconnection, the earthing arrangements for the installation and the impedences of the circuit concerned (413-02-04).

(g) circuit or equipment vulnerable to a typical test.

Note: For simple installations the foregoing information may be given in a schedule, if symbols are used they should conform to BS 3939.

Some British Standard Symbols BS 3939

⊕ Joint or junction box (example, 3 outlets).	⬜ Main control or intake point
✕ Lighting point	⬛ Main or sub-main switch
✕| Wall mounted lighting point	[t°] Thermostat
✗ Emergency or safety lighting point	ⵔ Bell
✗ Lighting point with switch	◷ Clock
⌒ Two pole, two way switch	▮ Watchman operated device or key operated switch
⅄ Socket outlet	$\frac{30A}{}$ Fuse
⅄³ Multiple socket outlet	⊡ Circuit breaker
⅄ Switched socket outlet	Isolator

The purpose of providing diagrams, charts and tables for an installation is so that it can be inspected and tested in accordance with Chapter 71 of the Regulations and to provide any new owner of the premises (should the property change hands) with the fullest possible information concerning the electrical installation.

It is essential that diagrams, charts and tables are kept up to date.

Typical charts and diagrams for a small installation are illustrated below.

Schedule of installation at ....

Type of circuit	Points served	Phase Conductor mm^2	Protective Conductor mm^2	Protective devices	Type of wiring
Lighting	10 downstairs	1 mm^2	1 mm^2	5 Amp Type 2 MCB	PVC/PVC
Lighting	8 upstairs	1 mm^2	1 mm^2	5 Amp Type 2 MCB	PVC/PVC
Immersion Heater	Landing	2.5 mm^2	1.5 mm^2	15 Amp Type 2 MCB	PVC/PVC
Ring	10 downstairs	2.5mm^2	1.5 mm^2	30 Amp Type 2 MCB	PVC/PVC
Ring	8 upstairs	2.5 mm^2	1.5 mm^2	30 Amp Type 2 MCB	PVC/PVC
Shower	Bathroom	6 mm^2	2.5 mm^2	30 Amp Type 2 MCB	PVC/PVC

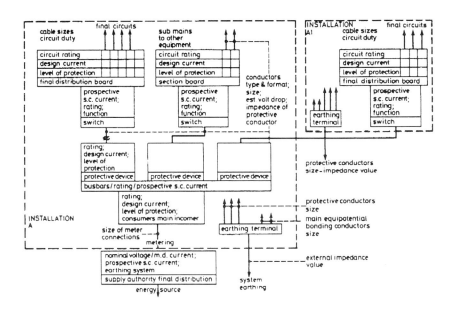

Warning Notices *(514-10)*

A warning notice stating the maximum voltage present should be fixed to every item of equipment (or enclosure) which contains circuits operating at voltages in excess of 250 volts and where the presence of such a voltage would not normally be expected.

Where accessories, control gear or switchgear are wired on different phases of a 3 phase supply, but can be reached simultaneously, a notice must be placed in a position where anyone removing an accessory, or gaining access to the terminals of control gear, switchgear etc. is warned of the maximum voltage.

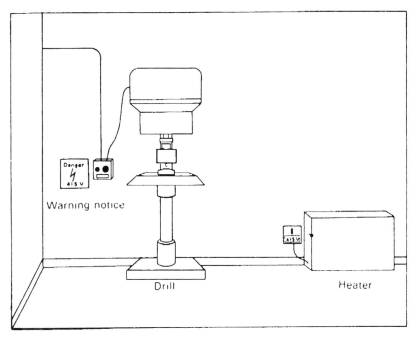

Inspection and Testing *(514-12)*

Upon completion of an electrical installation the electrical contractor should fix in a prominent position on or near the main distribution board of the installation, a label with details of the date of the last inspection and the recommended date of the next inspection.

The notice must be inscribed with characters (not smaller than 11 point), as illustrated below.

'IMPORTANT'

This installation should be periodically inspected and tested, and a report on its condition obtained, as prescribed in the Regulations for Electrical Installations issued by the Institution of Electrical Engineers.

Date of last inspection ....

Recommended date of next inspection ....

✳ Residual Current Device - Notices *(514-12-02)*

When an installation incorporates a residual current device, a notice must be fixed in a prominent position, at or near the main distribution board. It should be printed in indelible characters, not less than 11 point in size and should read as follows:

> This installation, or part of it, is protected by a device which automatically switches off the supply if an earth fault develops. Test quarterly by pressing the button marked 'T' or 'Test'. The device should switch off the supply and should then be switched on to restore the supply. If the device does not switch off the supply when the button is pressed, seek expert advice.

Caravans *(608-07-05)*

When electrical circuits have been installed in a touring caravan a notice (as illustrated) made of durable materials must be fixed on or near the main switch in the caravan.

> ### INSTRUCTIONS FOR ELECTRICITY SUPPLY
>
> **TO CONNECT**
>
> 1. Before connecting the caravan installation to the mains supply, check that:
>   - (a) the supply available at the caravan pitch supply point is suitable for the caravan electrical installations appliances, and
>   - (b) the caravan main switch is in the OFF position.
>
> 2. Open the cover to the appliance inlet provided at the caravan supply point and insert the connector of the supply flexible cable.
>
> 3. Raise the cover of the electricity outlet provided on the pitch supply point and insert the plug of the supply cable.
>
> **THE CARAVAN SUPPLY FLEXIBLE CABLE MUST BE FULLY UN-COILED TO AVOID DAMAGE BY OVERHEATING.**
>
> 4. Switch on at the caravan mains switch.
>
> 5. Check the operation of residual current devices, if any, fitted in the caravan by depressing the test button.
>
> **IN CASE OF DOUBT OR, IF AFTER CARRYING OUT THE ABOVE PROCEDURE THE SUPPLY DOES NOT BECOME AVAILABLE, OR IF THE SUPPLY FAILS, CONSULT THE CARAVAN PARK OPERATOR OR HIS AGENT OR A QUALIFIED ELECTRICIAN.**
>
> **TO DISCONNECT**
>
> 6. Switch off at the caravan main isolating switch, switch off at the pitch supply point and unplug both ends of the cable.
>
> **PERIODIC INSPECTION**
>
> Preferably not less than once every three years and more frequently if the vehicle is used more than normal average mileage for such vehicles, the caravan electrical installation and supply cable should be inspected and tested and a report of their condition obtained as prescribed in the Regulations for Electrical Installations published by the Institution of Electrical Engineers.

Earthing and Bonding

A warning notice (as illustrated) must be fitted in a visible position near to the point of connection of an earthing conductor to an earth electrode, or a bonding conductor to an extraneous conductive part.

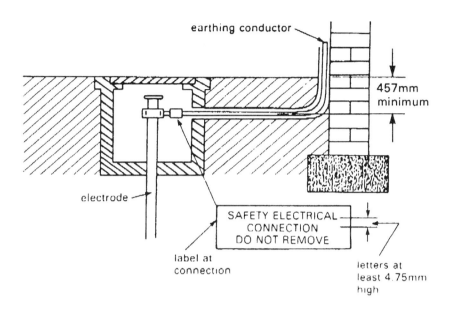

earthing conductor

457mm minimum

electrode

SAFETY ELECTRICAL
CONNECTION
DO NOT REMOVE

label at connection

letters at least 4.75mm high

[✱] Where protection against indirect contact is achieved by earth-free local equipotential bonding covered by Regulations 471-11/12 a notice, indelibly printed in letters not less than 4.75 mm high should be attached *(514-13-02)* and should read as follows:

> The equipotential protective bonding conductors associated with the electrical installation in this location MUST NOT BE CONNECTED TO EARTH.
>
> Equipment having exposed-conductive-parts connected to earth must not be brought into this location.

17

Inspection and Testing (Part 7)

✳ Introduction

Detailed methods of testing are no longer contained in the appendices of the 16th Edition. In order to provide a more comprehensive approach to testing, reference should be made to the IEE On-Site Guide and the IEE Guidance Notes No. 3 Inspection and Testing.

General

Inspection should comprise careful scrutiny of the installation, supplemented by testing to;

1. Verify safety of persons and livestock.

2. Verify protection against damage to property by fire and heat.

3. Establish that the installation is not damaged and has not deteriorated.

4. Identify installation defects or non-compliance with regulations.

Note: Attention is drawn to HSE Guidance Note GS38, 'Electrical test equipment for use by electricians' published by HMSO, which advises on the selection and safe use of suitable test probes, leads, lamps, voltage indicating devices and other measuring equipment.

Initial Inspection

Reasons for Inspection and Testing

The purpose of inspection of electrical installations is to verify that installations are safe and comply with the requirements of Regulations.

Test Instruments

Test instruments should be regularly checked and re-calibrated to ensure accuracy. The serial number of the instrument used should be recorded with test results, to avoid unnecessary re-testing if one of a number of instruments is found to be inaccurate.

For operation, use and care of test instruments, refer to manufacturer's handbook.

✳ General *(711)*

During its installation or on completion, every installation must be inspected and tested before being connected to the supply and energised. This should be done in such a manner that no danger to persons or damage to property or equipment can occur, even if the circuit tested is defective.

The following information should be made available to the persons carrying out the inspection and testing of an installation.

- Diagrams, charts or tables indicating:
  - (a) the type of circuits
  - (b) the number of points installed,
  - (c) the number and size of conductor,
  - (d) the type of wiring system

- The location and types of devices used for:-
  - − protection
  - − isolation and switching

- Details of the characteristics of the protection devices for automatic disconnection, the earthing arrangements for the installation, the impedances of the circuits and a description of the method used.

- Details of circuits or equipment sensitive to tests - e.g. central heating controls with electronic timers and displays

Note: Information may be given in a schedule for simple installations. See example below for a domestic installation. A durable copy of the schedule relating to distribution board must be provided inside or adjacent to the distribution board.

Schedule of installation at ....

Type of circuit	Points served	Phase Conductor mm^2	Protective Conductor mm^2	Protective devices	Type of wiring
Lighting	10 downstairs	1 mm^2	1 mm^2	5 Amp Type 2 MCB	PVC/PVC
Lighting	8 upstairs	1 mm^2	1 mm^2	5 Amp Type 2 MCB	PVC/PVC
Immersion Heater	Landing	2.5 mm^2	1.5 mm^2	15 Amp Type 2 MCB	PVC/PVC
Ring	10 downstairs	2.5mm^2	1.5 mm^2	30 Amp Type 2 MCB	PVC/PVC
Ring	8 upstairs	2.5 mm^2	1.5 mm^2	30 Amp Type 2 MCB	PVC/PVC
Shower	Bathroom	6 mm^2	2.5 mm^2	30 Amp Type 2 MCB	PVC/PVC

17/2

Inspection *(712)*

 A detailed inspection should be made of installed electrical equipment, usually with the part of the installation being inspected disconnected from the supply. The inspection should verify that it:

- complies with the British Standards or harmonised European Standards (this may be ascertained by mark or by certificate furnished by the installer or manufacturer)

- is correctly selected and erected in accordance with these Regulations

- is not visibly damaged so as to impair safety

The detailed inspection must include the following where relevant:

- Connections of conductors
- Identification of conductors
- Routing of cables in safe zones or mechanical protection methods
- Selection of conductors for current-carrying capacity and voltage drop
- Connection of single-pole devices for protection or switching in phase conductors only
- Correct connection of socket outlets and lampholders
- Presence of fire bariers and protection against thermal effects
- Methods of protection against direct contact (including measurements of distances where appropriate) i.e.
    - protection by insulation of live parts
    - protection by barriers or enclosures
    - protection by obtacles
    - protection by placing out of reach
- Methods of protection against indirect contact:
    - presence of protective conductors
    - presence of earthing conductors
    - presence of supplementary equipotential bonding conductors
    - earthing arrangements for combined protective and functional purposes
    - use of Class II equipment or equivalent insulation
    - non-conducting location (including measurement of distances, where appropriate)
    - earth-free local equipotential bonding
    - electrical separation

- Prevention of mutual detrimental influence

- Presence of appropriate devices for isolation and switching

- Choice and setting of protective and monitoring devices

- Labelling of circuits, fuses, switches and terminals

- Selection of equipment and protective measures appropriate to external influences

- Presence of undervoltage protective devices

- Adequency of access to switchgear and equipment

- Presence of danger notices and other warning notices

- Presence of diagrams, instructions and similar information

- Erection methods

Note: During any re-inspection of an installation all pertinent items in the check list should be covered.

✳ Testing *(713)*

The following items, (where relevant to the installation being tested), must be tested in the following sequence:

- Continuity of protective conductors including main and supplementary bonding

- Continuity of ring circuit conductors

- Insulation resistance

- Site applied insulation

- Protection by separation of circuits

- Protection by barriers or enclosures provided during erection

- Insulation of non-conducting floors and walls

- Polarity

- Earth loop impedance

- Earth electrode resistance

- Operation of residual durrent devices

Suitable reference methods of testing are described in the guide notes on the Regulations. The use of other methods of testing is not precluded provided that they will give results which are no less effective.

If a test indicates failure to comply, that test, and the preceding tests (whose results may have been affected by the fault) must be repeated after rectification of the fault.

Continuity of Protective Conductors *(713-02)*

The initial tests applied to protective conductors are intended to verify that the conductors are both correctly connected and electrically sound, and also the resistance is such that the overall earth fault loop impedance of the circuits is of a suitable value to allow the circuit to be disconnected from the supply in the event of an earth fault, (within the disconnection times selected to meet the requirements of Regulation 413-02-09).

Every protective conductor, including main bonding conductors and supplementary bonding conductors, should be tested to verify that the conductors are electrically sound and correctly connected.

Use a low resistance ohmeter for these tests.

Method 1

Step 1

Connect one terminal of the continuity tester to a long test lead and connect this to the consumer's earth terminal, as illustrated.

Step 2

Connect the other terminal of the continuity tester to a short lead and use this to make contact with the protective conductor at various points on the installation, testing such items as switch boxes and socket outlets.

The resistance reading obtained by the above method actually includes the resistance of the test leads. Therefore the resistance values of the test leads should be measured and this value deducted from any resistance reading obtained for the installation under test.

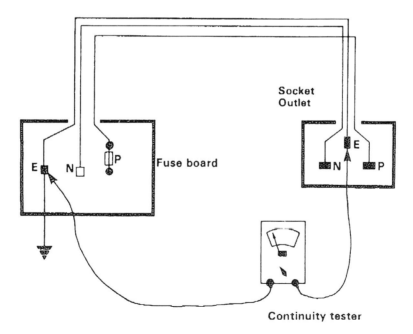

If the distance between the fuseboard and circuit under test involves the use of very long test leads, an alternative (Method 2) using the phase conductors as a test lead may be used.

- Strap the phase conductor to the protective conductor at a distant socket outlet, so as to include all of the circuit and test between phase and earth terminals at the fuseboard as illustrated

- The resistance measured by the above method includes the resistance of the phase conductor from the main switch to the point under test

The approximate resistance of this conductor can be obtained by joining together the phase and neutral conductors at the socket outlet (at the point under test) and measuring the resistance as shown below. The value of conductor resistance is half the value obtained by this test.

The value of earth continuity conductor resistance is calculated as the initial reading, minus phase conductor resistance.

Method 2 (cont'd)

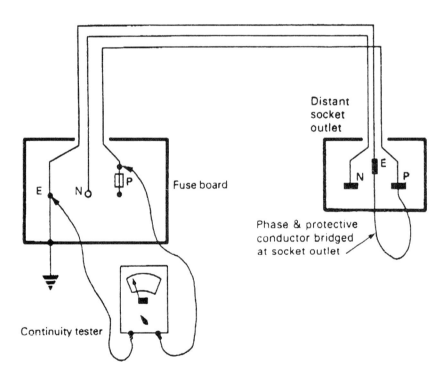

Alternative to Method 2

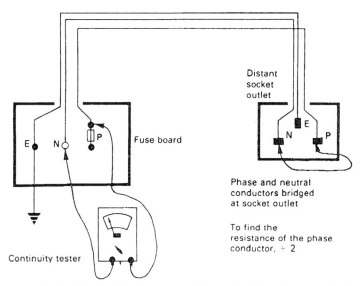

The test methods detailed below, as well as checking the continuity of the protective conductor, also provide a measure of $(R_1 + R_2)$ which enables the designer to verify the calculated earth-fault loop impedance Zs.

A low-resistance continuity tester should be used for these tests.

Earth Continuity and $R_1 + R_2$ Valve

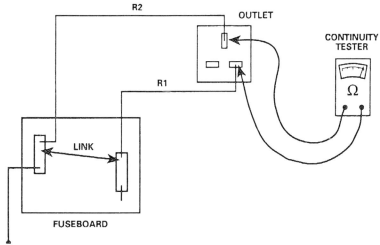

Note: Follow a similar procedure to Method 2, remembering to find resistance of the phase conductor ÷ 2.

✳ Testing of protective conductors comprising steel enclosures

When the protective conductor of the installation is a steel conduit or trunking, the following test procedure should be observed.

- carry out the continuity test with a low resistance ohmmeter using the method previously described

- visually inspect the steel enclosure to verify it is complete throughout the installation

If there are any grounds to suspect the soundness of the steel conduit or trunking installation forming the protective conductor, this should be verified using a phase earth loop impedance tester. (The test is described later in this section.

Should this test not confirm the soundness of the protective conductor to your satisfaction, the following test may be carried out.

Tests may be made using an a.c. or d.c. source of supply not exceeding 50 volts and with a current approaching 1.5 times the design current of the circuit, (but need not exceed 25 A).

When an a.c. test is used the current should be at the supply frequency.

When a d.c. test is used the protective conductor should be inspected throughout its length to verify that no inductor has been incorporated in the circuit, (e.g. operating coils and transformer windings, for instance if an earth monitoring unit has been incorrectly wired).

Note: This high current test could give rise to dangerous situations and should only be used where no other method of testing is available.

✳ Continuity of Ring Final Circuits Conductors (713-03)

A test must be made to verify the continuity of all line and protective conductors to every final ring circuit.

A number of test methods can be used. Some of these are explained below. The tests are to establish that the ring has not been interconnected to create an apparently continuous ring circuit where an actual break exists, as illustrated below. *One core illustrated for clarity.*

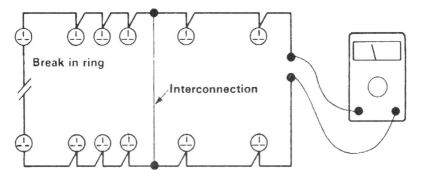

The Test

This test must be made to verify the continuity of phase neutral and protective conductors of every final ring circuit. The test result should also establish that the ring has not been interconnected to create an apparent continuous ring circuit which is actually broken. See illustration.

Using a low-resistance ohmeter, follow the method shown

Test 1: Ring Continuity

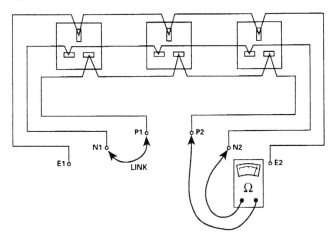

Mark up ends of ring circuit conductors E1, P1, N1, E2, P2, N2. Link P1 to N1 and measure resistance between P2 and N2. Note reading Repeat test, linking E1 to P1 and note reading.

Test 2: Socket Outlets

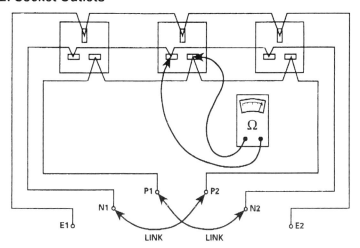

Link together N1 to P2, N2 to P1 and test each socket outlet, noting readings. The readings at each socket outlet should be substantially the same value. Repeat test, linking E1 to P2, E2 to P1 and note readings. This test may also be used for obtaining values of (R1 + R2). See Test 3.

Test 3: Ring Continuity R$_1$ + R$_2$ Value

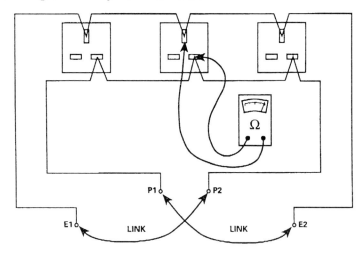

Visual inspection of ring circuit conductors

An alternative to the above methods for verifying that no interconnection multiple loops have been made in a ring circuit is for the installer to inspect each conductor throughout its entire length.

Insulation Resistance *(713-04)*

These tests are to verify that the insulation of conductors and electrical accessories and equipment is satisfactory and that electrical conductors or protective conductor are not short circuited, or showing a low resistance (which would indicate a deterioration in the insulation of the conductors).

Type of Test Instrument

An insulation resistance tester should be used which is capable of providing a d.c. voltage of not less than twice the nominal voltage of the circuit to be tested (r.m.s. value for an a.c. supply). The test voltage need not exceed the values below when loaded with 1mA.

- – 500V d.c. for installations connected to 500V.

- – 1000V d.c. for installations connected to supplies in excess of 500V and up to 1000V.

Pre-test checks

Ensure that neons and capacitors are disconnected from circuits to avoid inaccurate test value being obtained.

Disconnect control equipment or apparatus constructed with semi-conductor devices. These devices will be liable to damage if exposed to the high test voltages used in insulation resistance tests. This requirement will include certain types of rcd.

Ensure lamps and current using equipment are disconnected, and all fuses, switches and MCB's closed.

Circuit voltage:	Up to 500 volts	500 - 1000 volts	Between SELV and LV
Test voltage DC:	500 v	1000 v	500 v
Miniumum insulation resistance:	0.5 MΩ	1.0 MΩ	5.0 MΩ

See Table 71A of the IEE Regulations

Insulation resistance tests to earth

All fuses should be in, switches and circuit breakers closed, where practicable any lamps removed, appliances and fixed equipment disconnected. The phase and neutral conductors are connected together at the distribution board and a test is made as illustrated using an insulation resistance tester with test leads being connected between joint phase and neutral conductors and earth.

The reading obtained should not be less than

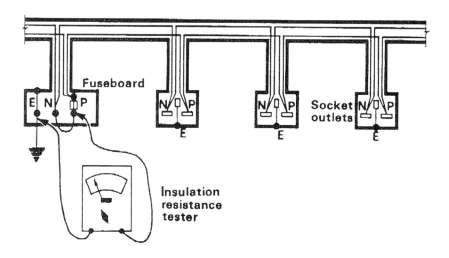

The above test cannot be carried out on TN-C (earth concentric) systems because the earth and neutral are common.

Insulation resistance tests between poles

All fuses should be in, switches and circuit breakers closed, where practicable any lamps removed, appliances and fixed equipment disconnected. For single phase circuits the test leads are connected between the phase and neutral conductors in the distribution board.

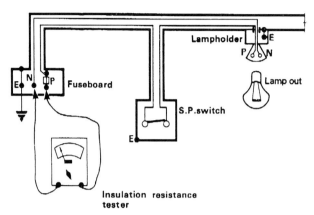

For three phase and neutral systems the following four tests are required as illustrated. The insulation resistance value should not be less than 0.5 megohms.

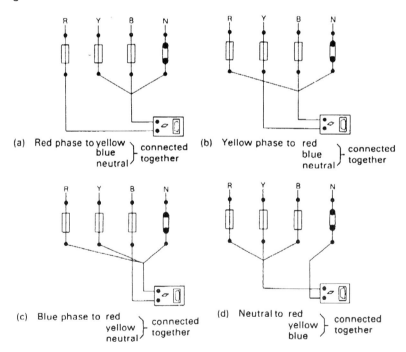

(a) Red phase to yellow, blue, neutral — connected together

(b) Yellow phase to red, blue, neutral — connected together

(c) Blue phase to red, yellow, neutral — connected together

(d) Neutral to red, yellow, blue — connected together

Note: Where any circuits contain two-way switching the two-way switches will require to be operated and another insulation resistance test carried out, including the strapping wire which was not previously included in the test.

Equipment

When fixed equipment such as cookers have been disconnected to allow insulation resistance tests to be carried out, the equipment itself must be insulation resistance tested between live points and exposed conductive parts.

The test results should comply with the appropriate British Standards. If none, the insulation resistance should not be less than 0.5 megohms.

Testing Class 2 Equipment

- Detailed physical examination
- Insulation test 500V d.c.
- Flash test

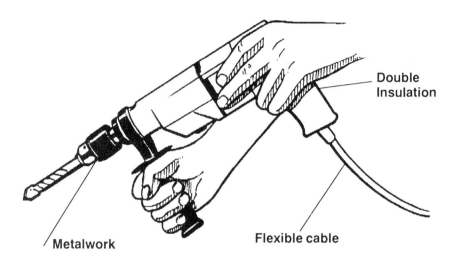

Double Insulation

Metalwork

Flexible cable

The electrical test performed is the "flash" test. British Standards require a test voltage across basic insulation of 1250V and across the supplementary insulation of 2500V resulting in a combined voltage of 3750V. The same voltage is required if the two separate layers are replaced by a single layer of reinforced insulation. It is generally accepted that a voltage of 3 kV is acceptable when repeated periodic testing is performed.

NOTE: Caution is essential in employing this method. Double-insulated equipment frequently incorporates electronic speed control which can be damaged by flash testing. Manufacturer's advice should be sought before applying the test.

The test may only be carried out by a competent person.

✳ Site Applied Insulation *(713-05)*

When protection against direct contact is afforded by insulation which has been applied to live parts of equipment during erection on site, a test should be made to ensure that the insulation is capable of withstanding an applied test voltage equivalent to that specified in the British Standard for similar type tested equipment.

Where protection against indirect contact is provided by supplementary insulation applied to equipment during erection, satisfactory insulation must be verified.

- The insulating enclosure affords a degree of protection not less than IP 2X (BS 5490 classifications). See note below

- The insulating enclosure is capable of withstanding without breakdown or flashover an applied test voltage equivalent to that specified in the British Standards for similar factory-built equipment.

These tests should be regarded as being of a special nature and would not usually be carried out in the testing and inspection of a completed installation.

Note: Busbar chambers, switch fuses and switchboards of unit factory-built construction are not considered to be 'site-built' assemblies.

Electrical Separation of Circuits *(713-06)*

Where safety extra low voltage (SELV) circuits are installed an inspection of the installation should be made to verify the separation of the separated circuits and/or continuity tests made to ensure that electrical separation has been achieved.

Protection by Barriers or Enclosures *(713-07)*

Where barriers and enclosures have been installed to prevent contact with conducting parts, they must comply with BS 5490 classification IP 2X and IP 4X.

Note: IP 2X - No contact can be made with a probe greater than 12 mm diameter and less than 80 mm long.

IP 4X - No contact can be made with any object of 1 mm thickness or 1 mm diameter.

Insulation of Non-Conducting Floors and Walls *(713-08)*

When protection against indirect contact is provided by a non-conducting location, such as an all-insulated room in a testing laboratory of a works, the resistance of the floors and walls of the room to the main protective conductor of the electrical installation should be measured at not less than three positions on each relevant surface.

One of the measured points must be between 1 m and 1.2 m from an extraneous conductive part (such as a water pipe or metal window frame).

 The test results must comply with regulations 413-04-07 and insulating arrangements must be able to withstand a test voltage of at least 2 kV and not pass a leakage current exceeding 1 mA in normal conditions of use.

Note: The regulations do not specify the test method or type of instrument *to be used.*

Such tests should be regarded as being special in character requiring the advice of those experienced in this field, i.e. probably the person who will use the non-conductive location.

Polarity *(713-09)*

This test must be carried out to verify that:

(a)   All fuses, circuit-breakers and single pole control devices such as switches are connected in the phase conductor only.

(b)   The centre contact of an Edison-type screw lampholder is connected to the phase conductor and the outer metal threaded parts are connected to the neutral or earthed conductor.

(c)   Any socket outlets have been correctly installed, i.e. phase pin of 13A socket outlet on right when viewed from the front.

The installation must be tested with all switches in the 'on' position and all lamps and power consuming equipment removed.

A test of polarity can be carried out using a continuity tester as illustrated.

Polarity Test Lighting

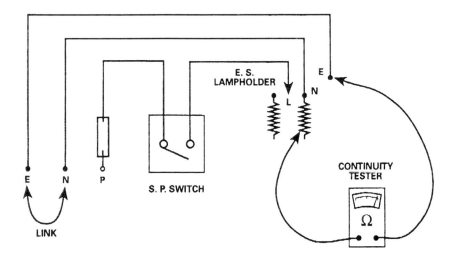

Polarity Test Socket Outlet

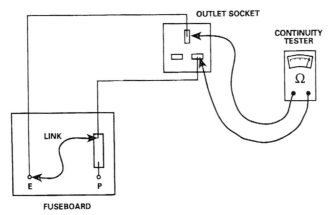

For ring circuits - if the ring circuit continuity test has been carried out to Regulation 713-03 polarity will have been confirmed during the test.

Earth Fault Loop Impedance *(713-10)*

The earth fault current loop comprises the following parts, starting at the point of fault for a phase to earth loop.

> Circuit protective conductor; the main earthing terminal and earthing conductors; for TN systems the metallic return path (or in the case of TT and IT systems the earth return path); and the path through the earth and neutral point of the transformer, the transformer winding and the phase conductor from the transformer to the point of fault.

The earth fault loop of the TN-S system is illustrated

Note: The impedance of the earth fault loop is denoted by (Zs)

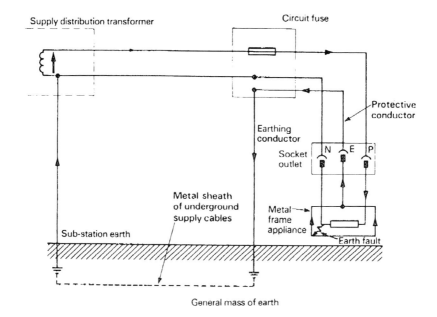

Earth fault loop impedance testing

The most common method of testing the earth fault loop impedance is by using a phase earth loop impedance tester.

Phase/Earth Loop Tests

Phase/earth loop tests can be made using the following methods:

(a) Using an instrument which connects a resistor of known value from the phase conductor to the protective conductor. A fault current of around 20 amperes is circulated through the earth fault loop and R for about 30 to 50 milli-seconds and is either measured directly on an ammeter in series with R or by using a voltmeter to measure the p.d. across R.
The value of loop impedance (Zs) being

$$\frac{\text{the supply voltage}}{\text{fault current}} \qquad Zs = \frac{U_o}{I_f}$$

The value of Zs above includes resistance (R) and the value of R must be subtracted from the value of Zs to give the true value of loop impedance. This is usually incorporated within the instrument, the scale being compensated to indicate a direct reading with the value of the internal resistance subtracted.

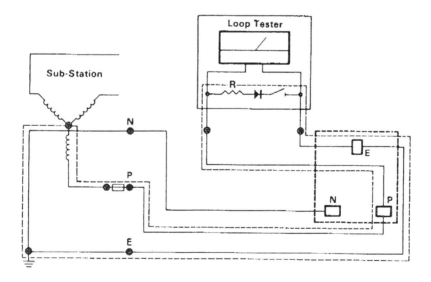

Note: In using such a test instrument, care must be taken to ensure that no ill effects can arise in the event of any defect in the earthing circuit such as would arise if there was a break in the protective conductor of the system under the test. This would prevent the test current from flowing and the whole of the protective conductor system would be connected directly to the phase conductor.

(b) Using a test instrument with alternating current at 10 amperes or less or using rapidly reversed direct current.

Where the protective conductor is steel conduit, metal trunking or the outer sheath of SWA cable, two tests have to be made if using test instrument as in (b), above.

(i) A test at the most distant position of the installation with all equipotential bonding conductors in place (Zs Ω).

(ii) A test at the mains position with main equipotential bonding conductors disconnected (ZE Ω).

The loop impedance value being twice the value of test (i) less the value of test (ii).

Earth Electrode Resistance *(713-11)*

After an earth electrode has been installed it is necessary to verify that the resistance of the electrode does not raise the earth fault loop impedance to an undesirable level.

There are two common methods of testing earth electrode resistance

- Alternating current method; employing a mains supply transformer; connected as illustrated

- Proprietary earth testers

Alternating Current Method

One of the difficulties in using this method is that the effect of back EMFs due to electrolytic action have to be taken into account. In addition, there is the possibility of stray currents being present in the soil due to leakages from the distribution system.

If the tests are made at power frequency, the source of current used for the test must be isolated from the mains supply (by using a double-wound transformer) and the electrode under test disconnected from all sources used for testing.

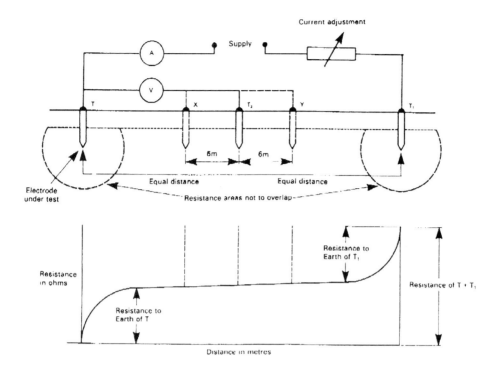

An alternating current of a steady value is passed between the earth electrode under test (T) and an auxiliary earth electrode (T_1) placed at such a distance from (T) that the resistance areas of the two electrodes do not overlap. A second auxiliary earth electrode (T_2), (which may be a metal spike driven into the ground) is then inserted half-way between (T) and (T_1), and the voltage drop between (T) and (T_2) measured. The resistance of the earth electrodes is then the voltage between (T) and (T_2) divided by the current flowing between (T) and (T_1) provided that there is no overlap of the resistance.

To check that the resistance of the earth electrodes is a true value two further readings are taken with the second auxiliary electrode (T_2) moved 6 m further from, and 6 m nearer to (T) respectively. If the three results are substantially in agreement, the mean of the three readings is taken as the resistance of the earth electrode under test (T). If there is no such agreement the tests are repeated with the distance between (T) and (T_1) increased.

The test is made with either current at power frequency, in which case the resistance of the voltmeter used must be high (of the order of 200 ohms per volt), or with alternating current from an earth tester comprising a hand-driven generator, a rectifier (where necessary), and a direct-reading ohmeter.

Proprietary Earth Testers

These have a d.c. hand-operated generator fitted with a current reverser to provide an a.c. supply at the output terminal or an a.c. hand-operated generator; or they may be battery operated.

Proprietary earth testers pass a current through the resistance under test from C1 to C2, and the resultant potential is measured between P1 and P2 causing the galvanometer to deflect.

WARNING The use of rubber gloves is recommended as a precaution against accidental high potentials in the soil.

Effect of Stray Currents

If stray currents are present in the soil this will cause the instrument pointer to oscillate. To eliminate this effect, the speed at which the generator is being driven must be either increased or decreased.

Measuring earth electrode resistance with a megger earth tester

The link between C1 and P1 should be as short as possible or its resistance will be included in the result. Alternatively, the leads from C1 and P1 should be the same length, in order to be of equal resistance so as to balance the Wheatstone Bridge.

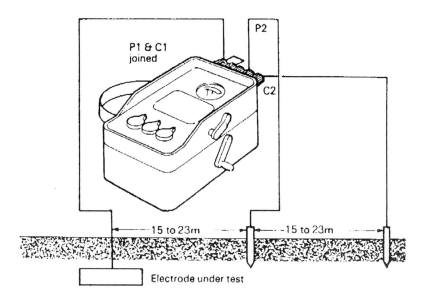

Operation of Residual Current Devices
(BS 4293) *(713-12)*

Residual current devices should be tested by simulating appropriate fault conditions, using a test meter.

The test is made on the load side of the circuit breaker, between the phase conductor of the circuit protected and the associated circuit protective conductor, so that a suitable residual current flows. All loads normally supplied through the circuit breaker are disconnected during the test.

The rated tripping current shall cause the circuit breaker to open within 0.2s or at any delay time declared by the manufacturer of the device.

> *NOTE: When the circuit breaker has a rated tripping current not exceeding 30mA and has been installed to reduce the risk associated with direct contact as indicated in the Note to Regulation 412-06-02, a residual current of 150mA should cause the circuit breaker to open within 40ms.*

In no event should the test current be applied for a period exceeding one second.

The effectiveness of the test button or other test facility which is an integral part of the circuit breaker should also be tested.

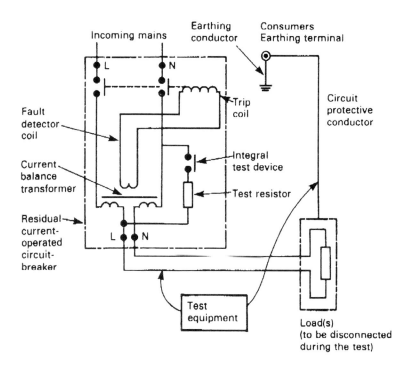

When making an earth fault loop impedance test on an installation fitted with an rcd this may trip out, especially where the tripping time of the rcd is 30 milli-seconds or less. In these circumstances, the rcd should be temporarily short-circuited to permit the earth loop impedance test to be carried out (remembering to remove the short-circuit when this test is completed). Since the rcd is inoperative during this time, care should be taken to see that no other tests (or other uses of the circuit) are undertaken until the rcd is back in circuit. Warning notices, etc. should be displayed.

Alterations to Installation *(721)*

When changes in the use of buildings occur either by change of ownership or use, alterations or additions to the electrical installations often take place. It is important that the person carrying out the electrical work ensures that the work complies with the IEE Regulations and that the existing electrical installation will function correctly and safely.

A completion certificate must be made out and issued for all the work involved in the alteration and any defect found in related parts of the existing installation reported to the person ordering the work by the electrical contractor (or a competent person).

Periodic Inspection and Testing *(731)*

Periodic inspection and testing of electrical installations and equipment must be carried out to ensure, as far as is reasonably practical;

- Safety of persons and livestock against electric shock and burns

- Protection of property from damage by fire and heat

- Compliance with the requirements of the Regulations.

This inspection should consist of careful scrutiny of the installation without dismantling, or by partial dismantling as required, supported by testing.

The periodic inspection and testing of installations is recommended on both the Completion and Inspection Certificate as follows:

	Not Exceeding
Domestic installations	10 years
Commercial installations	5 years
Industrial installations	3 years
Leisure complexes	1 year
Temporary installations on construction sites	3 months

The method of inspecting and test should be in accordance with the requirements of the Regulations. An Inspection Certificate must be completed and given to the client.

NOTE: Shorter intervals between inspections may be necessary depending on the nature of installation. Refer to designers specifications.

Reporting *(741)*

Following an inspection and test of an installation, a report must be provided by the person carrying out the inspection (or someone authorised to act on his behalf) to the person ordering the work.

Certificates

A combined Completion and Inspection Certificate must be completed and signed by a competent person. It must state that the installation has been:

- designed

- constructed

- inspected and tested

in accordance with the Regulations.

In the case of a small installation, compliance might be achieved by an electrician verifying all these items; a large installation carried out by a major contractor might require the signatures of three persons, i.e the designer, the electrican installing it, and the engineer carrying out the final inspection and test.

FORMS OF COMPLETION AND INSPECTION CERTIFICATE

(as prescribed in the IEE Regulations for Electrical Installations)

(1.) (see Notes overleaf)

DETAILS OF THE INSTALLATION

Client:

Address:

DESIGN

I/We being the person(s) responsible (as indicated by my/our signatures below) for the Design of the electrical installation, particulars of which are described on Page 3 of this form CERTIFY that the said work for which I/we have been responsible is to the best of my/our knowledge and belief in accordance with the Regulations for Electrical Installations published by the Institution of Electrical Engineers, 16th Edition, amended to (3.) (date . . . .) except for the departures, if any, stated in this Certificate.

The extent of liability of the signatory is limited to the work described above as the subject of this Certificate.

For the DESIGN of the installation:

Name (In Block Letters):   Position:

For and on behalf of:

Address:

(2.) Signature: (3.) Date

CONSTRUCTION

I/We being the person(s) responsible (as indicated by my/our signatures below) for the Construction of the electrical installation, particulars of which are described on Page 3 of this form CERTIFY that the said work for which I/we have been responsible is to the best of my/our knowledge and belief in accordance with the Regulations for Electrical Installations published by the Institution of Electrical Engineers, 16th Edition, amended to (3.) (date . . . .) except for the departures, if any, stated in this Certificate.

The extent of liability of the signatory is limited to the work described above as the subject of this Certificate.

For the CONSTRUCTION of the installation:

Name (In Block Letters):   Position:

For and on behalf of:

'dress:

(2.) Signature: (3.) Date:

INSPECTION AND TEST

I/We being the person(s) responsible (as indicated by my/our signatures below) for the Inspection and Test of the electrical installation, particulars of which are described on Page 3 of this form CERTIFY that the said work for which I/we have been responsible is to the best of my/our knowledge and belief in accordance with the Regulations for Electrical Installations published by the Institution of Electrical Engineers, 16th Edition, amended on (3.) (date . . . .) except for departures, if any, stated in this Certificate.

The extent of liability of the signatory is limited to the work described above as the subject of this Certificate.

For the INSPECTION AND TEST of the installation:

Name (In Block Letters):   Position:

For and on behalf of:

Address:

I RECOMMEND that this installation be further inspected and tested after an interval of not more than . . . . years. (5.)

(2.) Signature: (3.) Date:

(6) page 1 of . . . . pages

PARTICULARS OF THE INSTALLATION

(Delete or complete items as appropriate)

Type of Installation   New/alteration/addition/to existing installation

Type of Earthing (312-03):   TN-C   TN-S   TN-C-S   TT   IT
(Indicate in the box)
                     ☐   ☐   ☐   ☐   ☐

Earth Electrode: Resistance ....ohms

                     Method of Measurement ....

                     Type (542-02-01) and Location ....

Characteristics of the supply at the origin of the installation (313-01):

            Nominal voltage .... volts

            Frequency .... Hz   Number of phases ....

	ascertained	determined	measured
Prospective short-circuit current ....kA			
Earth fault loop impedance (Z_E) ....ohms			

            Maximum demand . . . . . . .A per phase

            Overcurrent protective device - Type BS ....   Rating . . . . . A

Main switch or circuit breaker (460-01-02): Type BS ....   Rating . . . . . A No of poles . . . . .

            (if an r.c.d., rated residual operating current $I_{\Delta n}$ ....mA.)

Method of protection against indirect contact:

1.   Earthed equipotential bonding and automatic disconnection of supply ☐

or

2.   Other ☐   (Describe) ....

Main equipotential bonding conductors (413-02-01/02, 546-02-01): Size............mm2

Schedule of Test Results: Continuation .... pages

Details of departures (if any) from the Wiring Regulations (120-04, 120-05).................

Comments on existing installation, where applicable (743-01-01): ....

(6) page 3 of . . . . pages

Reproduced from the IEE On-Site Guide with acknowledgements to the Institution of Electrical Engineers.

Insulation Resistance Testing

Check Sheet

1.	P1	—	P2
2.	P1	—	P3
3.	P2	—	P3
4.	P1	—	N
5.	P2	—	N
6.	P3	—	N
7.	P1	—	E
8.	P2	—	E
9.	P3	—	E
10.	N	—	E

Disconnect

Pilot lamps
Capacitors
Electronic devices
Fluorescent lamps
Current using equipment
Lamps

Connect

Rcd's closed
Fuses in
Switches closed
MCB closed
2-Ways operated

1 ph Tests 4-7-10

3 ph Tests 1-2-3-7-8-9

3 ph + N Tests 1 to 10

18

Projects

Note: Copies of manufacturer's cable data and certain British Standard data is required in order to complete Assignments C & D.

Assignment A

Installation at Corsby Village Hall *(see diagram page 18/4)*

Circuit to cooker control unit incorporating a 13A socket outlet.

Wiring system:

B.E. heavy gauge steel conduit incorporating four bends enclosing PVC insulated cables.

Information given:

TN-S system, 240 volts

Circuit design current I_b = 45A (data taken from module 11, example 3).

Overcurrent protective device - BS 1361 Cartridge fuse

Length of run = 14 m

No correction factors

External earth fault loop impedance Z_E = 0.4 Ω; k value for steel conduit = 47

Required to:

(a) Determine rating of overcurrent protective device, I_n

(b) Select cable size (from tables)

(c) Determine effective current carrying capacity of cable I_z

(d) Calculate voltage drop in cable and verify that circuit complies with voltage drop constraint

(e) Determine conduit size

(f) Determine maximum Z_s (from tables)

(g) Calculate actual Z_s and verify that circuit complies with shock protection constraint

(h)   Calculate earth fault current I_f

(i)   Determine disconnection time (from time/current characteristics)

(j)   Determine minumum c.s.a of circuit protection conductor and verify that circuit complies with thermal constraints.

(k)   Determine actual c.s.a. of conduit size to be used as c.p.c.

Assignment B

Installation at Corsby Village Hall *(see diagram page 18/4)*

Circuit to water heater in the kitchen; heater is 5 kW 240V instantaneous type.

Wiring system:

Heavy gauge PVC conduit incorporating six bends, with PVC insulated cables.

TN-S system, 240 volts

Information given:

Overcurrent protection device - BS 88 Part 2 HRC fuse

Length of run = 18 m

No correction factors

$Z_E = 0.4\ \Omega$

Required to:

(a)   Calculate design current, I_b

(b)   Determine rating of device, I_n

(c)   Select phase conductor size

(d)   Determine current carrying capacity of cable, I_z

(e)   Calculate voltage drop and verify circuit complies with voltage drop constraint.

(f)   Determine maximum Z_s

(g)   Calculate actual Z_s and verify circuit complies with shock protection constraint.

(h)   Calculate earth fault current I_f

(i) Determine disconnection time

(j) Determine c.p.c. size and verify that circuit complies with thermal constraint.

(k) Determine PVC conduit size.

Assignment C

Installation at Corsby Village Hall *(see diagram 18/4)*

Circuit to H & V control panel located in the boiler house.

Isolation device is 15A S P and N isolator.

Wiring system:

Heavy duty MICC cable with PVC sheath

TN-S system, 240V

Information given:

Design current, I_b = 11A

Overcurrent device is a BS 88 Part 2 HRC fuse

Length of run = 17.5 m

Ambient temperature in boiler house = 40°C

Z_E = 0.4 Ω

Required to:

(a) Determine rating of device, I_n

(b) Select cable size (from tables)

(c) Calculate voltage drop and verify circuit complies with voltage drop constraint.

(d) Determine maximum Z_S

(e) Calculate actual Z_S

(f) Calculate fault current, I_f

(g) Determine disconnection time

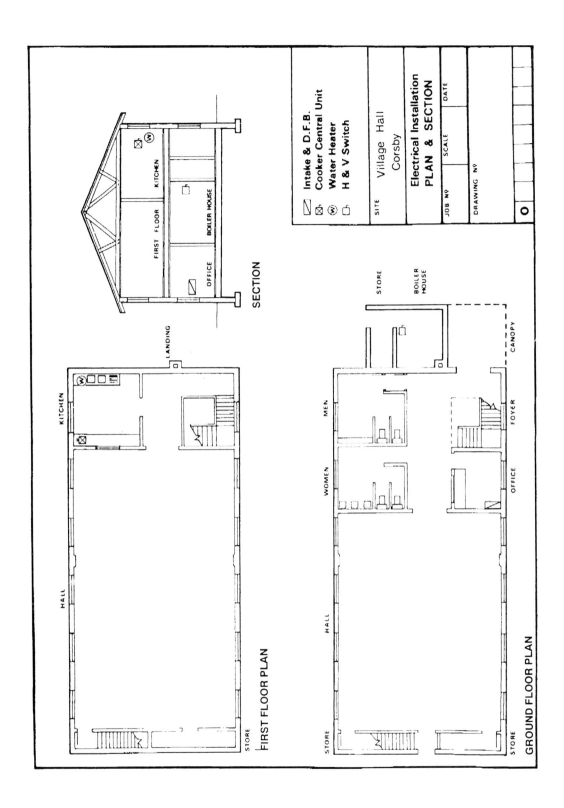

Intake & D.F.B.
Cooker Central Unit
Water Heater
H & V Switch

SITE Village Hall
Corsby

Electrical Installation
PLAN & SECTION

JOB Nº   SCALE   DATE

DRAWING Nº

SECTION

FIRST FLOOR PLAN

GROUND FLOOR PLAN

18/4

Assignment D

Installation to Area Lighting at Corsby Village Hall
(see diagram on following page)

Circuit to three luminaires in car park area.

Wiring system:

PVC SWA PVC cables laid in the ground and sanded and tiled.

PVC/PVC twin and cpc cables installed in column.

Information given:

> 3 No. 400 watt SON/T High Pressure Sodium lamps mounted on 5 m concrete columns each fitted with a SP & N fused cut out in the column base.
>
> Overcurrent protective device - BS 1361 Cartridge fuse
>
> Length of run = 126 m Depth of cable = 500 mm
>
> $Z_E = 0.4 \, \Omega$
>
> TN-S system, 240 volts
> Ground temperature 15°C (max) Air temperature 25°C (max)

Required to:

(a)   Calculate design current, I_b

(b)   Determine rating of device, I_n

(c)   Select cable size

(d)   Determine current carrying capacity of cables

(e)   Calculate voltage drop in cables and verify that circuit complies with voltage drop constraint.

(f)   Determine maximum Zs

(g)   Calculate actual Zs and verify that circuit complies with shock protection constraint.

(h)   Calculate earth fault current

(i)   Determine cable meets thermal constraints.

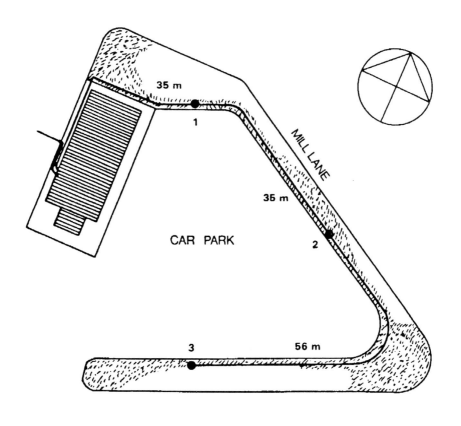

35 m

1

MILL LANE

35 m

2

CAR PARK

3   56 m

Client	CONSTRUCTION INDUSTRY TRAINING BOARD		
Site	VILLAGE HALL CORSBY		
Title	CONDUIT & CABLE LAYOUT AREA LIGHTING		
Job No.		Scale	Date

18/6

Model Answers

Assignment A

Information given:-

Supply - TN-S system,		240V
Design current Ib	=	45A
Protective device	=	BS 1361 cartridge fuse
Length of run	=	14 m
Z_E	=	0.4Ω
K for steel conduit	=	47 (from Table 54E, IEE Regulations)

No correction factors apply

(a)   For rating of protective device (I_n)

   From Regulation 433-02-01 $I_n \geq I_b$ = **45A**

(b)   For cable size

   From Regulation 433-02-01 $I_z \geq I_n$ = **45A**

   From Table 4D1A and 4D1B of the IEE Regulations (reference method 3) 10 mm^2 may be suitable (57A, 4.4mV)

(c)   Current carrying capacity of cable (I_z)

   From Regulation 433-02-01 $I_z \geq I_n \geq I_b$

   From Table 4D1A (reference method 3) of the IEE Regulations or 10 mm^2 I_z = **57A**

(d)   Voltage drop from Regulation 525-01-02 maximum permissible $V_d = 4\%$

$$= \frac{4}{100} \times 240 - 9.6V$$

   Actual voltage drop =

$$\frac{mV/A/m \times I_b \times length}{1000} = \frac{4.4 \times 45 \times 14}{1000} = \textbf{2.77V}$$

   Satisfactory as actual voltage drop (2.77V) < permissible voltage drop 9.6V.

(e) Determine conduit size

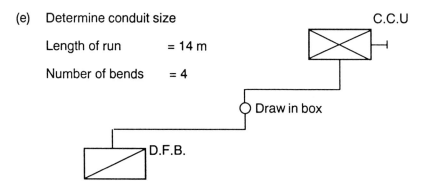

Length of run   = 14 m

Number of bends   = 4

As Table 5D in the IEE On-Site Guide, is limited to a 10 m length of run and to 2.5 m for runs incorporating four bends, choice is either:

(a) halve the run and consider circuit in two sections with intermediate draw in box or,

(b) carry out calculation using space factor formula

Method (a) preferable since the conduit run repeats itself, and the calculations are fewer and easier.

2 x 90° bends at each end, take length	=	7 m
From Table 5C, IEE On-Site Guide, cable term for 10 mm^2	=	105
for 2 conductors 105 x 2	=	210
From Table 5D, IEE On-Site Guide conduit size for 7 m run with two bends	**= 25 mm^2**	

(f) For maximum earth fault loop impedance (Z_S)

Regulation 413-02-09 states that disconnection should occur within 0.4 seconds as cooker control unit incorporates socket outlet.

From Table 41B1(b) for fuses to BS 1361

$$\text{maximum } Z_S \quad = \quad \mathbf{0.6\Omega}$$

(g) For actual Z_S

$$Z_S = Z_E + R_1 + R_2$$

$$Z_E = 0.4\Omega$$

From BS 4568 Part 1: 1981, for steel conduit should not exceed $R_2 \le 0.005\Omega/m$

From Table 6A, IEE On-Site Guide, R1 = 0.00183Ω/m for 10mm^2 phase conductor   x 1.38 from Table 6B, IEE On-Site Guide, = 0.00253Ω/m

$$\therefore \ Zs \ = \ 0.4 = (0.00253 \times 14) + (0.005 \times 14)$$

$$= 0.4 + 0.0354 + 0.07$$

$$= 0.5\Omega$$

Shock constraint satisfactory as actual Zs (0.5Ω) < maximum Zs (0.6Ω)

(h)   For earth fault current (If)

$$If \ = \ \frac{Uo}{Zs} \ = \ \frac{240}{0.5} \ = \ \textbf{480A}$$

(f)   For disconnection time (t)

From Fig 1, Appendix 3, IEE Regulations, t = **0.17 secs**

(j)   For thermal constraint

From Regulation 543-01-03, Table 54E, IEE Regulations, K for steel = 47

$$\therefore \text{cross-sectional area } S = \frac{\sqrt{I^2 t}}{K} = \frac{\sqrt{480^2 \times 0.17}}{47} = \frac{198}{47} = \textbf{4.2 mm}^2$$

(k)   Actual c.s.a. of c.p.c.

From page 10/9 of study notes, c.s.a. of 25 mm H.G. conduit = 131 mm^2

$$\therefore \ \textbf{25 mm}^2 \textbf{ conduit is suitable}$$

Assignment B

Information given:

Load - Water Heater	=	5kW
Supply - TN-S Uo	=	240V
Overcurrent protection	=	BS 88 Pt 2 HRC fuse
Length of run	=	18 m
ZE	=	0.4Ω

Wiring system is PVC conduit (6 bends) and PVC insulated cables

No correction factors apply

(a) For design current (Ib)

$$Ib = \frac{P}{Uo} = \frac{5 \times 1000}{240} = \mathbf{20.83A}$$

(b) Rating of protective device (In) select 25A size

Check In ≥ Ib

From Appendix 3, IEE Regulations, Fig. 3B, 25A BS 88 Pt 2 fuse should be available

∴ In = **25A**

(c) Phase conductor size

From Regulation 43-02-01 Iz ≥ In = **25A**

From Table 4D1A and B, IEE Regulations (reference method 3) **4 mm^2 may be suitable** (32A, 11mV)

(d) Current carrying capacity of cable (Iz)

From Regulation 433-02-01 Iz ≥ In ≥ Ib

From Table 4D1A, IEE Regulations (reference method 3) for 4 mm^2 Iz = **32A**

(e) Voltage drop

From Regulation 525-01-02 maximum permissible VD = 4%

$$= \frac{4}{100} \times 240 = 9.6V$$

Actual voltage drop

$$= \frac{mV/A/m \times Ib \times length}{1000} = \frac{11 \times 20.83 \times 18}{1000} = \mathbf{4.12V}$$

Satisfactory as actual voltage drop (4.12V) < permissible voltage drop 9.6V

(f) Maximum earth fault loop impedance (Zs)

Regulation 413-02-13 states that disconnection should occur within 5 seconds

From Table 41D(a), IEE Regulations for fuses to BS 88 Pt 2

Maximum Zs = **2.4Ω**

18/10

(g) For actual Zs

$$Z_S = Z_E + R_1 = R_2$$

$$Z_E = 0.4\Omega$$

From Table 6A, IEE On-Site Guide, $R_1 + R_2$ for 4 mm^2 phase conductor and 1.5 mm^2 c.p.c. is 0.01671Ω/m x1.38 from Table 6B, IEE On-Site Guide, $= 0.0231\Omega$/m

Actual Zs $= 0.4 + (0.0231 \times 18)$

$$= 0.82\Omega$$

Satisfactory as actual Zs $(0.82\Omega) <$ maximum Zs (2.4Ω)

(h) For earth fault current (I_f)

$$I_f = \frac{U_o}{Z_S} = \frac{240}{0.82} = \mathbf{292.68A}$$

(i) For disconnection time (t)

From Fig. 3B, Appendix 3, IEE Regulations, $t < 0.1$ seconds

(j) For thermal constraint

From Regulation 543-01-03, Table 54C, IEE Regulations, K for pvc insulated copper = 115

$$\therefore \text{cross sectional area S} = \frac{\sqrt{I^2 t}}{k} = \frac{\sqrt{292.7 \times 0.1}}{115} = \mathbf{0.804 \ mm^2}$$

Therefore 1.5 mm^2 c.p.c. does comply with thermal constraint

(k) For pvc conduit size

Consider dividing the circuit into 3 x 6 metre lengths incorporating two bends each

From Table 5C, IEE On-Site Guide, cable term for 4 mm^2 =   43

From Table 5C, IEE On-Site Guide, cable term for 1.5 mm^2 =   22

For 3 cables, (43 x 2) + (22 x 1)   =   108

From Table 5D, IEE On-Site Guide, conduit size for 6 m run with 2 bends, conduit term ≤ 108   =   **16 mm**

Note 16 mm conduit may not be a stock item so it is likely that 20 mm conduit would be used.

Assignment C

Information given:

Load - H & V control panel

Supply - TN-S system, 240V

Isolation - 15A SP & N isolator

Design current Ib	=	11A
Overcurrent protection	=	BS 88 Pt II H.R.C. fuse
Length of run	=	17.5 m
Ambient temperature	=	40°C
Z_E	=	0.4Ω

Wiring system is heavy duty pvc sheathed MICC cable.

(a) Rating of protective device (I_n)

$I_n \geq I_b = 11A$

From Appendix 3, Fig. 3B, IEE Regulations, 16A BS 88 Pt II fuse should be available

$\therefore I_n = $ **16A**

(b) From Regulation 433-02-01 $I_z \geq I_n = 16A$

From Table 4C1, IEE Regulations, for mineral cable (70°C sheath) Ca = 0.85

Correction for ambient temperature (40°C)

$$I_z \geq \frac{I_n}{Ca} = \frac{16}{0.85} = 18.82A$$

From Table 4J1A and 4J1B, IEE Regulations (reference method 1), **1 mm² may be suitable** (19.5A, 42mV)

(c) Voltage drop

From Regulation 525-01-02 maximum permissible Vd = 4%

$$= \frac{4}{100} \times 240 = 9.6V$$

Actual voltage drop

$$= \frac{mV/A/m \times I_b \times length}{1000} = \frac{42 \times 11 \times 17.5}{1000} = \textbf{8.09V}$$

(d) Maximum earth fault loop impedance (Zs)

Regulation 413-02-13 states that disconnection should occur within 5 seconds

From Table 41D(a), IEE Regulations, for fuses to BS 88 Pt 2

Maximum Z_S = **4.36Ω**

(e) For actual Zs

$$Z_S = Z_E + R_1 + R_2$$

From manufacturers data R_1 = 0.0179Ω/m, R_2 = 0.0019Ω/m

∴ Actual Z_S = 0.4 + (0.0179 x 17.5) + (0.0019 x 17.5)

Actual Z_S = **0.7465Ω**

(f) For earth current (If)

$$I_f = \frac{U_o}{Z_S} = \frac{240}{0.7465} = \textbf{321.5A}$$

(g) For disconnection time (t)

From Fig. 3B, Appendix 3, IEE Regulations, t = < **0.1 seconds**

Assignment D

Information given:

Load — 3 x 400W SON/T high pressure sodium lamps mounted on 5 m concrete columns each fitted with a SP & N fused cut-out in the base

Supply — TN-S system, 240V

Overcurrent protection	=	BS 1361 cartridge fuse
Length of run	=	126 m
Depth of cable	=	500 m
Z_E	=	0.4Ω
Ground temperature	=	15°C (max)
Ambient temperature	=	23°C (max)

Wiring system is PVC SWA cables laid in ground and sanded and tiled. PVC twin and cpc cables installed in column

(a)   For design current (I_b)

Total power consumed = 3 x 400 = 1200W

Table 1A, IEE On-Site Guide, states that specified lamp watts should be multiplied by 1.8 when assessing current demand, therefore:

design current I_b = $\dfrac{1200 \times 1.8}{240}$ = **9A**

(b)(i) Rating of protective device

$I_n \geq I_b$ = **9A**

From Appendix 3, Fig 1, IEE Regulations, 15A BS 1361 fuse should be available

∴ I_n = **15A**

  (ii) A suitable fuse for the cut-out in the lamp base from IEE Regulations, Apendix 3, Fig. 1, is **5A** BS 1361

(c)   For cable size

From Regulation 433-02-01 $I_t \geq I_n$ = 15A

From Table 4D4A and 4D4B (reference method 1) IEE Regulations, **1.5 mm^2 may be suitable** (21A, 29mV)

(d)   Current carrying capacity of cable (I_z)

From Regulation 433-02 $I_z \geq I_n \geq I_b$

From Table 4D4A (reference method 1) IEE Regulations for 1.5 mm^2, I_z = **21A**

(e)   Voltage drop

From Regulation 525-01-02 maximum permissible V_d = 4%

= $\dfrac{4}{100}$ x 240 = **9.6V**

Actual voltage drop to base of lamp 3 =

$\dfrac{mV/A/m \times I_b \times length}{1000}$ = $\dfrac{29 \times 9 \times 126}{1000}$ = **32.9V**

Unsuitable as actual voltage drop (32.9V) > permissible voltage drop 9.6V

This solution assumes that the current is the same in each section, which it is not. Re-consider the problem.

18/14

Consider the situation using 4 mm^2 cable which has a voltage drop of 11mV/A/m

DFB to base of L1 Vd $= \dfrac{mV/A/m \times Ib \times length}{1000}$

$= \dfrac{11 \times 9 \times 35}{1000} = 3.46V$

L1 to base of L2 $= \dfrac{11 \times 6 \times 35}{1000} = 2.31V$

L2 to base of L3 $= \dfrac{11 \times 3 \times 56}{1000} = 1.84V$

Total Vd from DFB to base of L3 = 7.61V

Note that this figure is only to the base of lamp 3; the voltage drop produced by the 5 m length of pole cable must be taken into consideration.

Individual lamp current $= \dfrac{400 \times 1.8}{240} = 3A$

From Table 4D2A (reference method 3), IEE Regulations, 1 mm^2 should be suitable (13A, 44mV)

Voltage drop at lamp (which should be added to the Vd at every lamp base)

$= \dfrac{mV/A/m \times Ib \times length}{1000} = = \dfrac{44 \times 3 \times 5}{1000} = 0.66V$

∴ Total VD from DFB to L3 $= 7.61V + 0.66V = \mathbf{8.27V}$

Satisfactory as total VD 8.27V < permissible Vd 9.6V

∴ 4 mm^2 PVC SWA cable and 1 mm^2 twin and cpc are satisfactory

(f) For maximum earth loop impedance (Z_S)

Regulation 471-08-03 states that disconnection should occur within 0.4 seconds (for fixed equipment outside the equipotential zone) as Table 41A, IEE Regulations

From Table 41B1(b) for 15A to BS 1361 fuse

Maximum Z_S = **3.43Ω**

(g) For actual Z_S

$$Z_S = Z_E + R_1 + R_2$$

$$Z_E = 0.4\Omega$$

Impedance value for PVC SWA PVC cable from BICC booklet 0.0161Ω/m

$R_1 + R_2$ value from Table 6A, IEE On-Site Guide, for pvc twin cable 1 mm^2

$$= 0.0362\Omega/m$$

From Table 6B, IEE On-Site Guide,
multiplier for pvc   =   1.38

Z_S to base of column 3   =   0.4 + (0.0161 x 126)

                                                =   2.42Ω

Z_S to lamp 3   =   2.42 + (0.0362 x 1.38 x 5)

Actual Z_S =   **2.67Ω**

Satisfactory as actual Z_S (2.67Ω) < maximum Z_S (3.43Ω)

(h) For earth current (I_f)

At base of column 3, $I_f = \dfrac{U_o}{Z_S} = \dfrac{240}{2.42} = $ 99.17A

From Appendix 3, Fig 1, disconnection time < 0.1 seconds

At lamp 3, $I_f \quad = \dfrac{U_o}{Z_S} = \dfrac{240}{2.67} = $ **89.88A**

From Appendix 3, Fig. 1, disconnection time for 5A BS 1361 fuse < 0.1seconds

Satisfactory as both disconnection times are less than 5 seconds *(611-02-04)*. Discrimination should be possible between fuses at lamp cut-outs and main fuse in DFB.

18/16

Procedure for the Selection of a Final Circuit Cable and Protective Device

See overleaf

STEP BY STEP PROCEDURE FOR THE SELECTION OF A FINAL CIRCUIT CABLE AND PROTECTIVE DEVICE

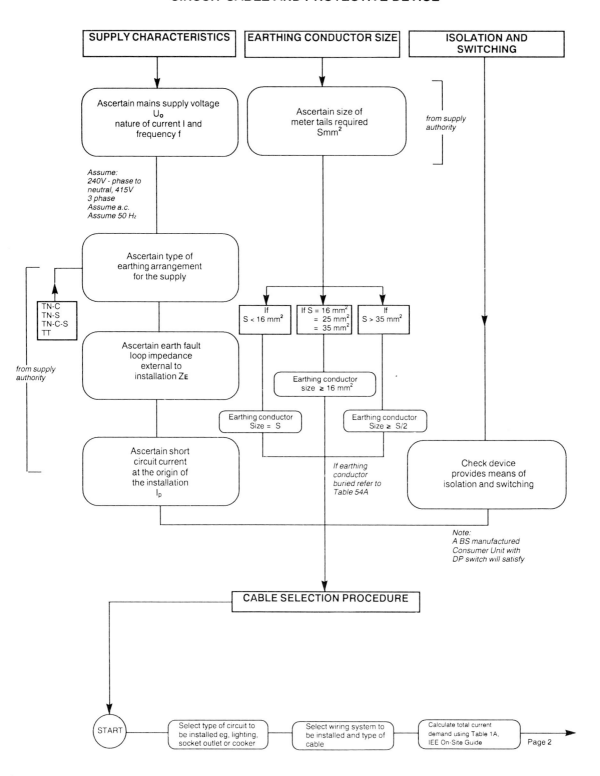

| SUPPLY CHARACTERISTICS | EARTHING CONDUCTOR SIZE | ISOLATION AND SWITCHING |

Ascertain mains supply voltage U_o nature of current I and frequency f

Ascertain size of meter tails required Smm^2

from supply authority

Assume:
240V - phase to neutral, 415V
3 phase
Assume a.c.
Assume 50 Hz

Ascertain type of earthing arrangement for the supply

TN-C
TN-S
TN-C-S
TT

from supply authority

If $S < 16\ mm^2$

If $S = 16\ mm^2$ $= 25\ mm^2$ $= 35\ mm^2$

If $S > 35\ mm^2$

Ascertain earth fault loop impedance external to installation Z_E

Earthing conductor size $\geq 16\ mm^2$

Earthing conductor Size = S

Earthing conductor Size $\geq$ S/2

Ascertain short circuit current at the origin of the installation I_p

If earthing conductor buried refer to Table 54A

Check device provides means of isolation and switching

Note:
A BS manufactured Consumer Unit with DP switch will satisfy

CABLE SELECTION PROCEDURE

START — Select type of circuit to be installed eg, lighting, socket outlet or cooker — Select wiring system to be installed and type of cable — Calculate total current demand using Table 1A, IEE On-Site Guide — Page 2

Ai

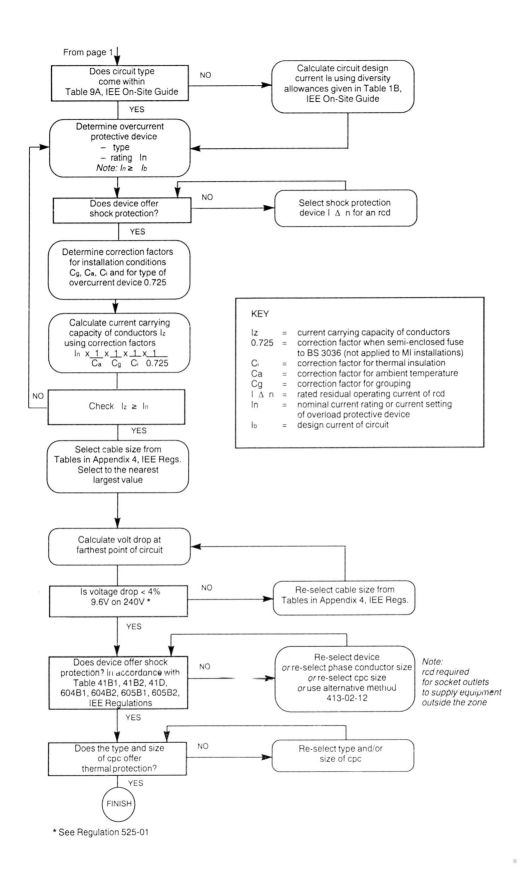

From page 1

Does circuit type come within Table 9A, IEE On-Site Guide — NO → **Calculate circuit design current I_b using diversity allowances given in Table 1B, IEE On-Site Guide**

YES ↓

Determine overcurrent protective device
- type
- rating I_n
Note: $I_n \geq I_b$

↓

Does device offer shock protection? — NO → **Select shock protection device $I \Delta n$ for an rcd**

YES ↓

Determine correction factors for installation conditions C_g, C_a, C_i and for type of overcurrent device 0.725

↓

Calculate current carrying capacity of conductors I_z using correction factors

$$I_n \times \frac{1}{C_a} \times \frac{1}{C_g} \times \frac{1}{C_i} \times \frac{1}{0.725}$$

↓

Check $I_z \geq I_n$ — NO (loops back)

YES ↓

Select cable size from Tables in Appendix 4, IEE Regs. Select to the nearest largest value

↓

Calculate volt drop at farthest point of circuit

↓

Is voltage drop < 4% 9.6V on 240V * — NO → **Re-select cable size from Tables in Appendix 4, IEE Regs.**

YES ↓

Does device offer shock protection? in accordance with Table 41B1, 41B2, 41D, 604B1, 604B2, 605B1, 605B2, IEE Regulations — NO → **Re-select device or re-select phase conductor size or re-select cpc size or use alternative method 413-02-12**

Note: rcd required for socket outlets to supply equipment outside the zone

YES ↓

Does the type and size of cpc offer thermal protection? — NO → **Re-select type and/or size of cpc**

YES ↓

FINISH

* See Regulation 525-01

KEY

I_z	=	current carrying capacity of conductors
0.725	=	correction factor when semi-enclosed fuse to BS 3036 (not applied to MI installations)
C_i	=	correction factor for thermal insulation
C_a	=	correction factor for ambient temperature
C_g	=	correction factor for grouping
$I \Delta n$	=	rated residual operating current of rcd
I_n	=	nominal current rating or current setting of overload protective device
I_b	=	design current of circuit

QUANTITY OR FACTOR	SYMBOL	REFERENCE	DATA CALCULATIONS	VALUE
Load current Design current	I I_b	Diversity tables 1A & 1B, IEE On-Site Guide.		
Protective device rating	I_n	Tables 41B1, 41B2, 41D, 604B1, 604B2, 605B1, 605B2 or time/current curves. Appx. 3, EE Regs.		
			Check $I_b \le I_n$	YES/NO
Cable size Grouping factor Ambient temperature Thermal insul. factor BS 3036 fuse factor Current carrying capacity	C_g C_a C_i 0.725 I_t	See attached notes Tables 4C1 and 4C2 Reg. 523-04 and Table 52A, IEE Regs. $I_t \ge \dfrac{I_n}{C_a \times C_g \times C_i \times 0.725}$		
Cable size chosen Tabulated current carrying capacity Effective current carrying capacity	I_t I_z	From Tables 4D1A - 4L4A, IEE Regs. $I_t \ge I_z$		
			Check $I_b \le I_n \le I_z$	YES/NO
Permissable V drop Actual V drop		Safe functioning of equipment (or 4%) Regs. 525-01 $\dfrac{mV/A/m \times I_b \times length}{1000}$		
		Check actual voltage drop $\le$ Permissable voltage drop		YES/NO

CABLE SELECTION PROCEDURE SHEET

QUANTITY OR FACTOR	SYMBOL	REFERENCE	DATA CALCULATIONS	VALUE
Shock Protection				
Maximum earth fault loop impedance	max Z_s	Table 41B1, 41B2, 41D, 604B1, 604B2, 605B1, 605B2, IEE Regs.		
External earth fault loop impedance	Z_E	Measure or obtain from electricity company		
c.p.c. size chosen		Select from BS 6004 initially		
Resistance of phase conductor	R_1	Table 6A and 6B, IEE On-Site Guide, manufacturers data or BS standards data		
Resistance of c.p.c.	R_2			
Actual earth fault loop impedance	Z_s	$Z_s = Z_E + (R_1 + R_2)$	Check $Z_s \leq$ Max Z_s	YES/NO
Thermal Constraint				
Fault current	I_f	$I_f = \dfrac{U_o}{Z_s}$ (240V phase to Earth)		
Time	t	Read time/current characteristics		
Factor for c.p.c.	k	Tables 54B to F, IEE Regs.		
Size c.p.c.	s	$\dfrac{\sqrt{I^2 t}}{k}$ mm^2	Check actual c.p.c. used is $\geq$ than s above	YES/NO

QUANTITY OR FACTOR	SYMBOL	REFERENCE	DATA CALCULATIONS	VALUE
For Groups Grouping correction factor Tabulated current carrying capacity required	C_g I_t	Table 4B1 and 4B2, IEE Regs. $I_t \geq \dfrac{I_n}{C_g \times (0.725)}$ (0.725 to be applied if BS 3036 fuse is used - not applicable to MI installations)		
Alternatively		Cable current carrying capacity should **not be less than the larger value** of I_t obtained after using BOTH of the following formulae provided that circuits of the group are **not** liable to simultaneous overload. $I_t \geq \dfrac{I_b}{C_g}$ and $I_t \geq \dfrac{\sqrt{(1.9)I_n{}^2 + 0.48 \times I_b{}^2 \ (1-C_g{}^2)}}{C_g{}^2}$ (1.9) to be applied only if BS 3036 fuse is used Any correction for ambient temperature or thermal insulation to be applied to I_t after these calculations have been completed i.e. min $I_t \geq \dfrac{I_t}{C_a \times C_i}$		

Av

INDEX

I